SAFE STORAGE
OF LABORATORY
CHEMICALS

SAFE STORAGE OF LABORATORY CHEMICALS

Edited by

DAVID A. PIPITONE

A Wiley-Interscience Publication

JOHN WILEY & SONS

New York   Chichester   Brisbane   Toronto   Singapore

Library of Congress Cataloging in Publication Data:

Main entry under title:

Safe storage of laboratory chemicals.

    Papers delivered at a symposium held in Kansas City
and sponsored by the Divisions of Small Business and
Chemical Health and Safety of the American Chemical
Society.
    "A Wiley-Interscience publication."
    Includes bibliographies and index.
    1. Chemical laboratories—Safety measures—Congresses.
2. Chemicals—Storage—Safety measures—Congresses.
I. Pipitone, David A. (David Andrew), 1952–
II. American Chemical Society. Division of Small
Business.   III. American Chemical Society. Division of
Chemical Health and Safety.

QD51.S22   1984   542'.028'9   83-21641
ISBN 0-471-89610-1

Printed in the United States of America

10 9 8 7 6 5 4

CONTRIBUTORS

JOHN BEQUETTE, University of Missouri, Columbia, Missouri
LESLIE BRETHERICK, Chemical Safety Matters, Berkshire, England
F. L. CHLAD, University of Akron, Akron, Ohio
J. GERLOVICH, Iowa Department of Public Instruction, Des Moines, Iowa
E. LAMAR HOUSTON, University of Georgia, Athens, Georgia
ALLEN G. MACENSKI, Hughes Aircraft Company, El Segundo, California
L. JEWEL NICHOLLS, University of Illinois at Chicago, Chicago, Illinois
DAVID A. PIPITONE, Lab Safety Supply Company, Janesville, Wisconsin
PATRICIA ANN REDDEN, Saint Peter's College, Jersey City, New Jersey
NORMAN V. STEERE, Norman V. Steere & Associates, Inc., Minneapolis,
Minnesota

To Helen and Cecilia,
whose love for chemistry
is nurtured through safety.

FOREWORD

The remarkably good record of the chemical industry in protecting its workers against on-the-job hazards is widely known. It is less widely appreciated that this reported low incidence of accidents is based largely on the favorable experience in our major chemical companies, where truly effective safety programs have been in operation for many years and company policies continue to strongly support good safety.

More chemical professionals work for small companies than for large, however, and in many of these small companies there has been a low degree of awareness of the potential hazards in handling chemicals. These companies have lacked safety specialists who are alert to dangerous situations and knowledgeable about preventive measures. As a result, accidents have occurred more frequently in the small companies (and these have not always been reported). Likewise, the safety practices in academic institutions generally have been far poorer than those in the well-run large chemical companies. Teachers have become increasingly aware of their deficiencies in the protection of students and staff, but they have felt helpless in the face of uncertainties about what should be done.

The papers that have been collected in this book are especially aimed at providing useful information for the newly safety conscious personnel of universities, colleges, and secondary schools and of small commercial operations. Concerns about the proper handling of chemicals in the stockroom and in bulk storage will be greatly relieved by using this book as a reference. Current and practical commentary is offered on relevant government regulations and on economic considerations essential for getting the most out of available dollars while ensuring compatibility with safety requirements.

These papers were delivered at a well-attended symposium in Kansas City sponsored by the American Chemical Society's Divisions of Small

Business and of Chemical Health and Safety. The number of beneficiaries
will now be greatly expanded by this publication. We are properly grateful
to everyone involved in the project.

MALCOLM RENFREW
Professor of Chemistry, Emeritus

University of Idaho
Moscow, Idaho
January 1984

PREFACE

Chemical health and safety is one of the primary concerns in today's workplace. No longer is chemical exposure just an environmental issue. Rather, with the rise of state "right-to-know" legislation, the proper storage and handling of chemicals, coupled with employer responsibility to provide accurate information, is becoming a legal issue in the workplace. This book is devoted to the proper storage of chemicals found in the storerooms and laboratories of academic, medical, and industrial institutions.

The chapters that comprise this volume were originally delivered as papers at a symposium sponsored by the American Chemical Society in September 1982. Publication as a book has crystallized the authors' thoughts and experience, providing a groundwork for those implementing safety procedures in both laboratories and storerooms.

Each chapter plays an integral role in the total management of chemicals. Knowledge for the identification of chemical storage hazards is abundantly detailed. Practical solutions that minimize or eliminate the potential for danger in storage are discussed thoroughly. Management policies for maintaining storeroom integrity are outlined.

This book is divided into two basic parts: storage guidelines and actual case histories. Storage guidelines include applications for flammable, corrosive, water-reactive, explosive, and other types of incompatible chemicals; precautionary labeling procedures; spill control measures; and computer selection for the retrieval of storage and chemical health information. The case histories include survey results, a creative approach to chemical storage at a major university, description of a computer database system for a large-scale warehouse and inventory concern, and a state plan for the disposal of hazardous chemicals from schools.

The information presented on the following pages will be useful to teachers, laboratory and safety directors, administrators, and consultants who desire a safer chemical workplace. Industrial hygienists and those conducting safety inspections may easily apply this information to assess chemical storage areas. Finally, engineers and architects who design new storage and laboratory facilities (or who revitalize current facilities) will benefit from understanding the necessary safety principles.

My sincere thanks and appreciation go to the individual authors for their fine work. I salute Chris Marlowe, together with the Divisions of Chemical Health and Safety and Small Business of the American Chemical Society, for sponsoring the original symposium on chemical storage. Thanks are due to the duPont company for a financial contribution to the symposium and to the J. T. Baker Chemical Company for permission to publish information on their SAF-T-DATA labeling system. I am grateful to Don Hedberg for introduction to laboratory safety, to Joel and Cele Pipitone, and to the numerous friends and relatives who, through good wishes and encouragement, have provided moral support. Special thanks are given to Jim Smith of John Wiley & Sons, Bess Hetrick for her help with typing of the manuscript, and my wife, Cheryl, for her patient support and friendship.

DAVID A. PIPITONE

Janesville, Wisconsin
January 1984

CONTENTS

xiii

APPENDIXES

SAFE STORAGE
OF LABORATORY
CHEMICALS

PART 1

STORAGE GUIDELINES

CHAPTER 1

STORAGE REQUIREMENTS FOR FLAMMABLE AND HAZARDOUS CHEMICALS

NORMAN V. STEERE

**Laboratory Safety and Design Consultant
Norman V. Steere & Associates, Inc.
Minneapolis, Minnesota**

Storage of chemicals should minimize safety and health hazards to personnel, equipment, buildings, and the environment. Safe storage will require appropriate construction, equipment, and operating practices. Chemical containers should be stored and ventilated to prevent breakage or leakage that might endanger someone who enters or works in the storage area or might cause deterioration of the chemicals or damage to containers, equipment, or the building. The storage facility should be separated and protected so that a fire or spill in the storage area is not likely to spread beyond the storage area.

Part of the requirements for safe storage of hazardous chemicals have been learned from adverse experience and incorporated into codes and regulations. Codes and regulations and code-enforcing authorities can provide some important guidance, but they usually cannot address all the safety and health aspects of chemical storage, particularly as new chemicals come into more common use and new hazards are discovered.

This chapter describes and discusses the safety requirements for storage of chemicals within a separate room within a building used for teaching, research testing, or other laboratory activities. The primary emphasis is on the storage requirements specified in the standards used by local or state fire authorities, insurance inspectors, and the Occupational Safety and Health Administration (OSHA) and other federal agencies.

Several National Fire Protection Association standards will be referred to:

Flammable and Combustible Liquids Code, NFPA 30-1981

Hazardous Chemicals Data, NFPA 49-1975

Code for Storage of Liquid and Solid Oxidizing Materials, NFPA 43A-1980

Code for Storage of Gaseous Oxidizing Materials, NFPA 43C-1980

Code for Storage of Pesticides in Portable Containers, NFPA 43D-1980

Fire Protection Standard for Laboratories Using Chemicals, NFPA 45-1982

Safety Standard for Laboratories in Hospitals, NFPA 56C-1980

Storage of ammonium nitrate in quantities of 1000 lb or more is governed by the Code for Storage of Ammonium Nitrate, NFPA 49-1980. Storage of cylinders of acetylene, fuel gases, hydrogen, and oxygen may be regulated by NFPA codes, depending on the sizes of containers and quantities stored.

1.1 GENERAL RECOMMENDATIONS

Safe storage of chemicals must begin with identification of the chemicals to be stored and their hazards. Since many chemicals have several hazards that may vary in the degree of severity, depending on quantity and concentration, it is often difficult to determine what protection is needed for safe storage and where best to store a particular chemical. For example, concentrated acetic acid is corrosive to the skin [at concentrations of 80% or greater it is defined as a "Corrosive Liquid" by Department of Transportation (DOT) regulations] and is also a combustible liquid with a flash point temperature of about 43°C. In contrast, vinegar, which has an acetic acid concentration of 4–8%, is considered both nontoxic and noncombustible.

When chemicals have multiple hazards it is important to store them in the most appropriate storage areas, and it may also be necessary to segregate them within that storage area (Figure 1.1). As another example, sulfuric acid and sodium hydroxide might be stored in a single storage area because they are both corrosive, but they would need to be protected from accidental mixing that could cause a hazardous chemical reaction. Separation, segregation, or isolation is recommended depending on the severity of hazard, total quantities stored, and the size and durability of individual containers.

The material and size of storage containers will affect the need for special storage practices and safety procedures. For example, if containers of flammable and combustible liquids are no larger than 5 gallons in size, there is no requirement for special provision to prevent liquid flow from the storage area into the adjoining building.

Ventilation is needed for chemicals and containers that may release dangerous or damaging quantities of vapors or gases that are flammable, corrosive, irritating, or toxic. Ventilation may also be needed for containers and chemicals that may produce annoying odors.

For every storage area there should be evacuation and emergency procedures to be followed in case of leak, spill, or fire within the room.

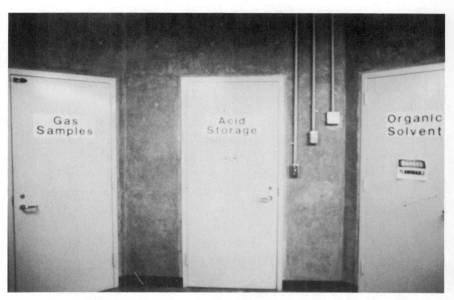

FIGURE 1.1   Separate storage rooms for gases, acids, and solvents.

Important safety and health considerations should include:

1. Ventilation needed for protection of health and prevention of corrosion.
2. Lighting needed for finding containers and reading labels.
3. Identification of storage locations.
4. Strength, stability, and corrosion resistance of shelving.
5. Aisles and storage arrangements that provide for safe access (this may require shallow shelves or pull-out trays or drawers and movable steps to reach storage above eye level).

Location of frequently used chemical storage should be based on consideration of safety in travel to the storage area and transport of chemicals to and from the storage area.

The cost of specialized storage space and the cost of time to obtain access to chemical storage facilities are basic considerations. Generally, the cost of space decreases as it becomes less specialized and farther from the point of use, but the time cost of obtaining chemicals increases as the travel distance increases. Chemicals stored at the bench or other work area should be those that are used frequently. Quantities should be limited to the minimum necessary, and the container size should be the minimum convenient.

Chemicals stored near the work area in cupboards, cabinets, and closets as backup supplies for the work should be limited to the minimum quantities necessary. The container size should be convenent. Chemicals in stockrooms or similar accessible supply areas should include all needed chemicals that are not stored in or near work areas. Quantities should be limited to the storage space provided. Chemicals stored in separate or detached areas should be limited to bulk quantities that cannot safely or economically be stored inside the building or stockroom areas.

1.1.1 Identification of Chemicals

Hazardous chemicals can be stored and handled more safely if they have labels that list precautions or can be used to refer easily and quickly to safety information. Although there has been no federal standard for precautionary labeling of laboratory chemicals, many chemical suppliers follow the voluntary labeling standard developed under procedures of the American National Standards Institute by the Manufacturing Chemists Association. A growing number of states have adopted legislation requiring precautionary labeling of chemicals. An example of the labeling system used by one company is described in Appendix 1.

Some labeling terms commonly used are "Combustible," "Flammable," "Corrosive," "Irritating," and "Toxic." Use of a single term will not always provide adequate information since many chemicals have more than one hazard. Another problem is that the term "Corrosive" includes materials that may be incompatible with each other. For example, strong mineral acids as well as strong alkaline materials are both corrosive, but if they are mixed accidentally in a storage area, they would react vigorously.

There are several systems and standards for labeling chemicals to communicate their hazards, but there is no uniformly accepted system for signaling hazards *and* conveying precautionary information. Each standard and system has advantages and limitations, which are described in the following sections. We believe that a combination of systems is needed so that labeling of laboratory reagents provides adequate safety and health information

1.1.1.1 Precautionary Labeling.

Many laboratory reagent containers are labeled by the manufacturer with the kind of precautionary information required for consumer products. Precautionary labeling for hazardous materials has been developed by the Chemical Manufacturers Association. Such precautionary labeling will contain the name of the chemical, a signal word such as "Warning" or "Danger," the key hazards such as "Flammable" or "Vapor Harmful," and statements of precautions to avoid the hazards (Figure 1.2).

FIGURE 1.2 Precautionary labeling on a laboratory safety can.

We believe that precautionary labeling is the minimum that should be required. For containers that are breakable and large enough to create a serious problem if hazardous contents are splashed or spilled, we recommend use of the NFPA hazard signal system to supplement precautionary labeling. Use of the hazard signal system is useful for emphasizing emergency hazards.

Some governmental jurisdictions (including several states and cities) have laws requiring precautionary labeling on all containers of hazardous materials. Such regulations also usually require the employer to have material safety data sheets available on their hazardous materials and to provide employees with training to recognize the hazards explained on the labels and in the data sheets. (This system was recommended for adoption as a national regulatory standard by NIOSH and an OSHA advisory commiteee.)

1.1.1.2 DOT Hazard Labeling System. The DOT hazard labeling system uses a color-coded diamond in which there is a symbol (such as a flame) and a term describing the major hazard of the material. Hazard classifications include "Flammable Liquid," "Flammable Gas," "Compressed Gas," "Corrosive," "Poison," "Radioactive," "Explosive," and "Oxidizer." Most chemicals are rated by what DOT considers the major single hazard, even though many chemicals have hazards in several categories and hazards of different degrees (Figure 1.3). Regulations of the DOT system are changing so that some chemicals are required to have two hazard labels.

Because many chemicals have hazards of different types and degrees, we consider the DOT system by itself as inadequate to communicate the storage and handling precautions that may be necessary for laboratory chemicals.

Department of Transportation regulations for labeling of hazardous materials in the transportation system must be used by laboratories that are packaging and shipping hazardous waste. In brief, the system requires specification containers and packaging, preparation of shipping papers that list hazardous materials first and in special terms, hazard labels on shipping containers and vehicles, and training of all employees who package and offer hazardous materials for disposal or other shipment.

1.1.1.3 NFPA 704 Hazard Signal System. The NFPA 704 hazard signal system, an NFPA standard, uses a color-coded diamond with four quad-

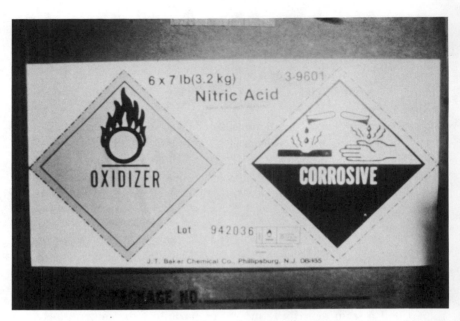

FIGURE 1.3 DOT hazard labels for oxidizer and corrosive on a shipping container of concentrated nitric acid.

rants in which numbers are used in the upper three quadrants to signal the degree of emergency health hazard, fire hazard, and instability/reactivity hazard. (The bottom quadrant can be used to indicate water reactivity, radioactivity, biohazards, or any other hazard.)

The three types of emergency hazard are signaled on a numerical scale of 4 to 0, with 4 = extreme hazard, 3 = severe hazard, 2 = moderate hazard, 1 = minor hazard, and 0 = no unusual hazard. Examples of the types of hazards and ratings are shown in Tables 1.1 and 1.2.

The NFPA hazard signal system is useful in alerting personnel to the degree of hazard of the chemical in a container and helpful in drawing attention to some of the storage needs and emergency equipment needed in case of spills or splashes. The limitations of the system are that chronic hazards are not signaled and that precautionary information is not given (Figure 1.4).

As an example of the need for precautionary labeling in addition to hazard signals, the emergency health hazard ratings are estimations of

TABLE 1.1   NFPA 704 Hazard Signal System—Hazard Ratings and Examples

Number	Emergency Health Hazard	Fire Hazard	Instability–Reactivity
4	**Extremely toxic**	**Flammable gas or Class IA liquid**	**Extremely shock sensitive and capable of detonation**
	Hydrogen cyanide	Hydrogen, methane ethyl ether, pentane	Picric acid dry benzoyl peroxide
3	**Toxic or Corrosive**	**Flammable liquid Class IB and IC**	**Shock-sensitive materials that may detonate under some conditions**
	Sodium cyanide, sulfuric acid	Acetone, methanol ethanol, toluene	Dilauroyl peroxide
2	**Moderately toxic**	**Combustible liquid Class II and IIIA**	**Unstable and water-reactive materials**
	Toluene, ether	Acetophenone	Sodium, sulfuric acid
1	**Irritating**	**Combustible, including Class IIIB liquids**	**Materials that may become unstable under heat or pressure**
	Acetone, MEK	Cod liver oil	Glacial acetic acid
0	**No unusual hazard**	**Noncombustible**	**Not reactive**

acute exposures by inhalation or skin contact and do not assess chronic exposures or hazards from ingestion or injection.

Hazard signal ratings are useful in assessing the need for precautions for storage, spills, splashes, and other emergency conditions involving major quantities of the concentrated material. Small quantities and diluted materials will usually present less health hazard, less flammability, and greater stability.

The hazard ratings that should generally be used in the hazard signal on a storage area are based on the hazard ratings of the material stored there. For example, the combined signal consists of the highest emergency health hazard rating, the highest fire hazard rating, and the highest instability/reactivity rating (Figure 1.5). The NFPA 704 standard does allow reduction of the hazard ratings on the storage area if materials with high hazards are stored in small quantities or dilute concentrations.

TABLE 1.2 Examples of Hazard Signals for Several Laboratory Reagents

Chemical	Emergency Hazard Signal		
	Health	Fire	Instability
Acetic acid	2	2	1
Acetone	1	3	0
Ethyl alcohol	0	3	0
Ethyl ether	1	4	1
Hydrochloric acid	3	0	0
Hydrogen cyanide	4	4	2
Picric acid	1	4	4
Sodium hydroxide	3	0	0
Sulfuric acid	3	0	2 + Water reactivity
Toluene	2	3	0
Xylenes	2	3	0

FIGURE 1.4 NFPA 704 hazard signals on laboratory safety cans.

FIGURE 1.5   NFPA 704 hazard signals identifying emergency hazards of chemicals stored in a storage building.

1.1.2   Separation and Isolation

Prevention of fire and other hazardous reactions in storage areas is based partly on avoiding unintentional mixing of chemicals from leaking or broken containers. The need for separation, segregation, or isolation will depend on the size and durability of storage containers, the potential for leakage, and the hazards of the chemicals (Figure 1.6).

FIGURE 1.6   Chemicals separated in storage by masonry walls.

"Separation" is defined by NFPA 49 as storage within the same fire area but separated by as much space as practicable or by intervening storage from incompatible materials. For example, NFPA 49 recommends that sulfuric acid be stored separate from combustible materials and that acetic acid be stored separate from oxidizing materials.

"Isolation" is defined by NFPA 49 as storage away from incompatible materials in a different storage room or in a separate and detached building located at a safe distance. As one example, NFPA recommends that storage of dry benzoyl peroxide (in industrial quantities) be isolated in well-detached, fire-resistant, cool, and well-ventilated buildings with no other materials stored therein.

"Segregated" storage is generally defined by NFPA standards as storage in the same room but physically separated by space from incompatible materials. NFPA 43A requires sills, curbs or intervening storage to maintain spacing. NFPA 43C specifies separation by at least 20 ft. (6.1 m).

1.1.3 Storage Containers

Large glass containers are susceptible to breakage in transportation and handling unless they are protected by shipping containers, carrying containers, or a heavy plastic coating (Figure 1.7).

Five-gallon metal containers are often too heavy and awkward for convenient and safe pouring of the contents. Fifty-five-gallon drums of solvents should generally not be allowed in laboratory work areas, but in special rooms equipped for storage or for dispensing.

FIGURE 1.7 Acid bottle protected from breakage by a bottle carrier. Bottle carrier also separates acid bottle from contact with containers of incompatible materials.

1.1.4   Ventilation

Ventilation is a fire-prevention requirement for storage rooms in which flammable liquids are stored or dispensed and is generally an occupational health recommendation for rooms in which such liquids are dispensed. Although not a code requirement, ventilation is certainly needed for any storage area in which there may be leakage of vapors or gases that are corrosive to metal or irritating, annoying or toxic to personnel.

1.1.5   Evacuation and Emergency Procedures

Emergencies that can occur within the storage area or that may affect the storage area include fire, spillage, release of radioactive material, release of gas, ventilation failure, escape of pathogens, and explosion.

Response to such emergencies will usually be limited to primary emergency procedures, unless the facility has provided additional special training for all personnel or has organized and equipped an emergency response team.

Emergency procedures are described in more detail later in this chapter.

1.2   STORAGE OF HAZARDOUS CHEMICALS

1.2.1   Combustible and Flammable Materials

Many organic and inorganic materials are combustible, and some have such a high degree of combustibility that they are designated as flammable. The most common requirement for special storage is for organic liquids that can release flammable concentrations of vapors at temperatures at or below 93.4°C (200°F).

Since code requirements for special storage are often stated in terms of the fire hazard classification of a material or in terms with a specific degree of hazard, it is necessary to describe the classifications and definitions in terms of flash-point temperatures:

Ignitable liquids, regulated by the Environmental Protection Agency (EPA) under the Resource Conservation and Recovery Act, include Class I flammable liquids and Class II combustible liquids.

Special storage is commonly required for quantities of flammable liq-

Classification	Term	Flash-Point Temperature
Class III	"Combustible Liquid"	Any flash point at or above 60°C
Class IIIB	"Combustible Liquid"	At or above 93.4°C (200°F)
Class IIIA	"Combustible Liquid"	Below 93.4°C (200°F) At or above 60°C (140°F)
Class II	"Combustible Liquid"	Below 60°C (140°F) At or above 37.8°C (100°F)
Class I	"Flammable Liquid"	Below 37.8°C (100°F)
Class IC	"Flammable Liquid"	Below 37.8°C (100°F) At or above 22.8°C (73°F)
Class IB	"Flammable Liquid"	Below 22.8°C (73°F) Boiling point at or above 37.8°C (100°F)
Class IA	"Flammable Liquid"	Below 22.8°C (73°F) Boiling point below 37.8°C (100°F)

uids and certain combustible liquids in excess of 120 gallons that cannot be stored in special wooden or metal storage cabinets. The liquids that require such special storage have flash-point temperatures at or below 93.4°C (200°F) and include all liquids identified as a flammable liquid, an ignitible liquid, or a combustible liquid in Class II or IIIA.

At least one approved liquid storage room is required within any health care facility regularly maintaining a reserve storage capacity in excess of 300 gallons (1135.5 liters).

Generally the only other combustible materials that require special storage are combustible gases and a limited number of materials classified as flammable solids. Storage requirements for combustible gases are discussed later in the section on compressed gases.

Flammable and Combustible Liquid Storage Rooms. The requirements for storage of flammable and combustible liquids are spelled out in OSHA Standards and in the NFPA Flammable and Combustible Liquids Code, NFPA No. 30-1981. The standards apply to storage of flammable liquids and Class II and IIIA combustible liquids [with flash-point temperatures at or below 93.4°C (200°F)].

Both sets of standards allow storage of significant quantities of such liquids within laboratory buildings in rooms that are specially separated from the rest of the building so that a spill or fire in the room is not likely to spread into the main building.

An inside storage room for flammable and combustible liquids will be reasonably safe and will meet or exceed both sets of standards if it meets the following requirements. Dispensing is discussed later in this chapter.

An inside storage room that does not exceed 150 square feet in floor area is permitted to contain up to 2 gallons per square foot of floor area in the room, if the room is separated from the building by construction having at least 1-hour fire resistance and all openings between the room and the building are protected by assemblies having a 1-hour fire-resistance rating. If it is desirable to increase the allowable storage capacity of such a room, the capacity can be increased to 5 gallons per square foot by providing the room with an automatic fire extinguishing system, which might be as simple as adding one or two automatic sprinkler heads.

An inside storage room needs to be ventilated to prevent possible accumulation of flammable concentration of vapors from container leaks or spills. Recommended ventilation is from floor level with a capacity of one cubic foot per minute of exhaust for each square foot of floor area in the room, with a minimum of 150 cubic feet per minute. (If there is dispensing in the room, there should be provision for ventilating the dispensing operations close to the points at which vapors are being emitted.)

If the containers of flammable and combustible liquids are no larger than 5 gallons in size, it should *not* be necessary to provide barriers to prevent spills in the room from spreading into the main building. If containers of Class I or II liquids *are* larger than 5 gallons in size, there is need for curbs, ramps, scuppers, or special drains, with a drainage capacity for all of the water that could be discharged from an automatic fire extinguishing system and from the fire hose streams that may be applied by the local fire department.

Wiring and electrical fixtures located in inside storage rooms must be suitable for the hazards. Explosionproof (Class I, Division 2) electrical equipment is required for prevention of explosions if flammable liquids (Class I) are being stored or dispensed. If only combustible liquids are being stored or dispensed, general-use wiring is acceptable.

If an inside storage room has an exterior wall, it will be classified by the 1981 edition of the NFPA Flammable and Combustible Liquids Code as a "cut-off room," for which there are two additional requirements: (a) exterior walls are required to provide ready accessibility for fire fighting; and (b) if Class IA or IB liquids are dispensed or if Class IA liquids are stored in containers larger than 1 gallon, the exterior wall or roof are required by NFPA 30-1981 to be designed to provide explosion venting to meet the requirements of NFPA 68-1978.

We do not believe there is a need for explosion venting in a small room used only for storage or in a room used for dispensing if adequate ventilation is provided.

Flammable liquids and combustible liquids can safely be grouped in a storage room or a storage cabinet that meets the requirements listed above for storage of flammable liquids. Some of the commonly encountered organic acids that are combustible materials and can appropriately be stored with flammable liquids include:

Acetic acid	Combustible liquid, Class II
Anthranilic acid	Combustible
Butyric acid	Combustible liquid, Class IIIA
Chloroacetic acid	Usually crystals; flash point 259°F; no label required unless liquid or solution
Citric acid	Combustible
Crotonic acid	Combustible liquid, Class IIIA
Oleic acid	Combustible liquid, Class IIIB
Oxalic acid	Organic acid that is corrosive to skin and highly toxic; DOT label is "Poison'
Stearic acid	Combustible liquid, Class IIIB
Toluenesulfonic acid	Combustible; DOT label is "Corrosive"

1.2.2 Oxidizers

Mineral acids, including those recognized as strong oxidizers such as nitric acid, perchloric acid, and sulfuric acid, should be separated from flammable and combustible materials. Such mineral acids should be stored in separate rooms, separate cabinets, or break-resistant containers if large glass bottles must be stored in proximity of combustible materials (Figure 1.8). For prevention of oxidization of wooden storage shelves (or corrosion of metal shelves), acid-resistant trays or mats should be provided under bottles of nitric, perchloric, and sulfuric acids.

The NFPA Code for Storage of Liquid and Solid Oxidizing Materials, NFPA 43A-1980, defines oxidizing material and four classes of oxidizer and establishes requirements based on quantities. Oxidizing material is defined as any solid or liquid that readily yields oxygen or other oxidizing gas or that readily reacts to oxidize combustible materials. The four classes of oxidizer are Class 1, Class 2, Class 3, and Class 4.

The primary hazard of Class 1 oxidizers is an increase of the burning rate of combustible material with which it comes in contact. Examples are 8–27.5% hydrogen peroxide solutions; magnesium perchlorate; nitric acid, 70% concentration or less; perchloric acid solutions, less than 60% by weight; and silver nitrate. The standard applies when quantities are stored in excess of 4000 lb (1816 kg).

Class 2 oxidizers are regulated when stored in quantities in excess of 1000 lb (454 kg). Class 2 oxidizers moderately increase the burning rate or may cause spontaneous ignition of combustible material with which it

FIGURE 1.8 Acids separated in storage by masonry walls.

comes in contact. Examples of Class 2 oxidizers are calcium hypochlorite, 50% or less by weight; chromic acid; hydrogen peroxide, 27.5–52% by weight; and sodium peroxide.

Class 3 oxidizers are regulated when stored in quantities in excess of 200 lb (91 kg). Class 3 oxidizers will cause a severe increase in the burning rate of combustible material with which it comes in contact or will undergo vigorous self-sustained decomposition when catalyzed or exposed to heat. Examples are ammonium dichromate, hydrogen peroxide, 52–91% by weight; perchloric acid solutions, 60–72.5%; and sodium chlorate.

Class 4 oxidizers are regulated when stored in quantities in excess of 10 lb (4.5 kg). Class 4 oxidizers can undergo an explosive reaction when catalyzed or exposed to heat, shock, or friction. Examples are ammonium perchlorate; ammonium permanganate; hydrogen peroxide, more than 91% by weight; perchloric acid solutions, more than 72.5%; and potassium superoxide.

The Code for Storage of Gaseous Oxidizing Materials, NFPA 43C-1980, applies to oxidizers in cylinders or other containers with an aggregate capacity in excess of 100 lb (45 kg) when in storage or connected to a manifold system. The standard applies to chlorine, chlorine trifluoride, fluorine, nitrous oxide, oxygen, and about 10 other gaseous oxidizing materials that are not readily available.

Emergency water may be needed in all of these storage areas since many of the chemicals are corrosive to human tissue.

Oxidizers shall be stored to avoid contact with incompatible materials such as ordinary combustibles, flammable liquids, greases, and other materials, including other oxidizers, that could react with the oxidizer or catalyze its decomposition.

If the class and quantity of oxidizer is regulated by NFPA standards, storage of oxidizers shall be segregated, cutoff or detached, depending on the class of oxidizer. The required fire-resistance for cutoff storage increases as the class increases. Class 4 oxidizers in regulated quantities are permitted to be stored only in detached storage. Class 3 oxidizers are permitted to be stored only on the ground floor of a building with no basement. Storage areas for Class 2 and 3 oxidizers in combustible containers shall be provided with means to vent fumes in a fire emergency; storage areas for Class 4 oxidizers shall be provided with means to vent fumes in any type of emergency.

Gaseous oxidizing materials are generally irritating, toxic and highly reactive chemically. They can react violently with finely divided metals and with organic and other materials that are readily oxidizable.

1.2.3   Corrosive and Irritating Chemicals

In addition to the oxidizers that are corrosive or irritating, alkalies and bases are corrosive or irritating. Those that are liquid in large glass containers, such as ammonium hydroxide, should be stored in a separate cabinet or separate area. Ventilation may be needed.

Although there are no code requirements for storage of corrosive or irritating chemicals, construction should limit spread of any liquid spills, and emergency water should be provided.

1.2.4   Toxic Chemicals

There are no code requirements for storage of toxic chemicals. However, ventilation should be provided if needed, as should emergency water in case of chemical splash. Construction should limit spread of any liquid spills.

Acid-sensitive materials such as the cyanides and sulfides should be stored in a location separate from acids or protected from contact with acids.

Pesticide storage shall be located or constructed so that runoff from fire-fighting operations will not contaminate streams, ponds, groundwater, land or buildings.

Storage areas for pesticides and other highly toxic chemicals should be secured when the storage areas are not supervised by a responsible person, so that the public is kept out.

1.2.5   Reactive and Incompatible Chemicals

If chemicals are to be stored that are reactive if exposed to air or water, they can safely be stored in sprinklered areas where sprinkler discharge would serve to prevent rupture of the outer container. Temperature control or refrigeration must be provided as needed for chemicals that deteriorate or react if their temperatures exceed safe limits recommended by the manufacturer or person synthesizing the chemical.

Although avoiding storage that may allow mixture of incompatible chemicals and hazardous chemical reactions under routine and emergency conditions is certainly desirable, published codes and guidelines cannot cover every chemical that is synthesized, formulated, or produced in quantity. Under some conditions, such as when chemicals are stored in small or break-resistant containers, code requirements may not be appropriate. Guidelines for segregation of incompatible chemicals may be found in NFPA 49-1975, Hazardous Chemicals Data; NFPA 491M-1975, Hazardous Chemical Reactions; and Coast Guard recommendations.

1.2.6   Compressed Gases

Combustible gases that are classified as fuel gases may be stored inside of a building up to a total gas capacity of 2500 cubic feet of acetylene or nonliquefied flammable gas or about 309 lb of propane or 375 lb of butane. If there is more than one storage area within a building, they must be separated by a distance of at least 100 feet. The quantity of acetylene or nonliquefied flammable gas in a storage area may be doubled if the storage area is protected with an automatic sprinkler system that will provide a density of at least 0.25 gallons per minute per square foot over an area of at least 3000 square feet. (Standard NFPA 51-1977 provides additional detail.)

Standard NFPA 50A-1978 establishes requirements for gaseous hydrogen systems having containers with a total content of 400 cubic feet or more. The standard also applies where single systems having a content of less than 400 cubic feet of hydrogen are located less than 5 feet from each other. Publications of the Compressed Gas Association give additional recommendations for storage.

1.2.7   Hazardous Waste

Hazardous waste must be classified into the appropriate hazard class for storage. It is recommended that separate storage areas be established for storage of waste to prevent damage to new material and to provide better inventory control for all material in storage.

1.3   EMERGENCY EQUIPMENT AND PROCEDURES

1.3.1   Fire Extinguishing

Automatic fire extinguishment and fire detection systems are to be connected to the facility fire alarm system and be arranged to immediately sound an alarm.

Fire extinguishers suitable for the particular hazards are to be located so that they will be readily available to personnel in accordance with NFPA 10, Standard for Portable Fire Extinguishers (Figure 1.9).

Information necessary for providing fire protection and other hazard control measures is sought by the requirement of the Fire Protection Standard for Laboratories Using Chemicals, NFPA 45-1982, that "When chemicals are ordered, steps shall be taken to determine the hazards and to transmit that information to those who will receive, store, use or dispose of the chemicals."

FIGURE 1.9   Fire extinguishers on a cart provide for mobile response to a variety of fires.

1.3.2   Emergency Water Needs and Requirements

Readily available supplies of water are needed for emergency flushing of chemicals off of personnel in all facilities and workplaces in which irritating, corrosive, or toxic chemicals are used. General requirements were set forth in OSHA Standards, and a recently adopted American National standard gives some specific guidelines. However, we believe that there are additional important needs that deserve consideration. These needs are described following a statement of the OSHA standards, NFPA 56C standards, and some details on the standard adopted by the American National Standards Institute (ANSI).

1.3.2.1   OSHA Standards for Emergency Water   The general OSHA requirement for emergency water is set forth in the section on Medical Services and First Aid in Subpart K of the OSHA Standards:

> Where the eyes or body of any person may be exposed to injurious corrosive materials, suitable facilities for quick drenching or flushing of the eyes and body shall be provided within the work area for immediate emergency use. [1910.151(c)]

In another section of the OSHA Standards there is a guideline for providing emergency water for small chemical splashes (in nonlaboratory areas):

> Near each tank containing a liquid which may burn, irritate, or otherwise be harmful to the skin if splashed upon the worker's body, there shall be a supply of clean cold water. The water pipe (carrying a pressure not exceeding 25 pounds) shall be provided with a quick opening valve and at least 48 inches of hose not smaller than three-fourths inch, so that no time may be lost in washing off liquids from the skin or clothing. Alternatively, deluge showers and eye flushes shall be provided in cases where harmful chemicals may be splashed on parts of the body. [1910.94(d), (9)(vii)]

There is a definite need to provide at least a 15-minute supply of water for flushing chemicals from the eyes, since many chemical manufacturers recommend on their product labels that the eyes should be washed for 15 minutes in case the chemical gets in the eyes.

1.3.2.2   NFPA 56C Standards for Emergency Water.   Standard NFPA 56C contains the following requirements:

> Where the eyes or body of any person may be exposed to injurious corrosive materials, suitable fixed facilities for quick drenching or flushing of the eyes and body shall be provided within the work area for immediate emergency use. Fixed eye baths shall be designed and installed to avoid injurious water pressure.
>
> If shutoff valves or stops are installed in the branch line leading to safety drenching equipment, the valves shall be OS and Y (outside stem and yoke), labeled for identification, and sealed in the open position.
>
> The installation of wall mounted portable eye wash stations shall not preclude the adherence to the provisions of this section.

1.3.2.3   ANSI Standard Z358.1-1981 for Emergency Eyewash and Shower Equipment.   The American National Standards Institute adopted a standard for emergency eyewash and shower equipment in June 1981. The standard was developed by the Emergency Shower and Eyewash group of the Industrial Safety Equipment Association under ANSI standard-making procedures.* The ANSI standard defines four different types of water delivery system and sets standards for performance, installation, test procedures, maintenance, and training.

Emergency Showers.   The standard for emergency showers calls for shower heads 82–96 inches above the floor with a spray pattern centered at least 16 inches from any obstruction and 20 inches in diameter at a height of 60 inches above the floor. The valve actuator is to be easily located and readily accessible, with a handle not more than 175.26 cm (69 inches) above the standing level, and the control valve is to remain on without holding it (Figure 1.10). The standard specifies that "emergency showers shall be accessible within 10 seconds and should be within a travel distance no greater than 30.5 m (100 feet) from the hazard."

The standard requires that emergency showers be capable of delivering a minimum of 113.6 liters per minute (30 gallons per minute) of water and that the shower be connected to a 1-inch IPS minimum water supply.

*This standard is available for a cost of $5.00 plus a handling fee of $2.00 from the Institute at 1430 Broadway, New York, New York 10018.

FIGURE 1.10   Emergency shower and separate eyewash in a storage room. (This arrangement does not permit one person to flush both eyes and body simultaneously.)

If a shower enclosure is provided, the standard specifies that there shall be a minimum unobstructed area of 86.4 cm (34 inches) in diameter.

The standard specifies that the shower be activated weekly to flush the line and verify proper operation.

The standard states that "All employees who might be exposed to chemical splashes shall be instructed in proper use of safety showers." The same requirement for instruction is included for eyewash equipment.

Plumbed and Self-Contained Eyewash Equipment. The ANSI standard specifies that "a means shall be provided to assure that a controlled flow of potable water or its equivalent is provided to both eyes simultaneously at a low enough velocity so as not to be injurious to the user." The specified volume of water is not less than 1.5 liters per minute (0.4 gallons per minute) for a period of 15 minutes.

The standard calls for the eyewash control valve to be simple to operate and able to turn on in one second or less and to be designed so that the water flow remains on without requiring the use of the operator's hands (Figure 1.11).

The standard for location of emergency eyewash units is the same as that for emergency showers, within 10 seconds and 100 feet.

Eye/Face Wash Equipment and Hand-Held Drench Hoses. The standard for eye/face wash units is similar to that for eyewash units, but the flow rate recommended is 11.4 liters per minute (3.0 gallons per minute).

Drench hoses are to be designed to provide a controlled flow of water to the eyes or a portion of the body at low enough velocity to avoid injury

FIGURE 1.11 Eye and face washer.

to the user, and to deliver water at a minimum rate of 11.4 liters/minute (3.0 gallons per minute).

Combinations of Shower and Eyewash or Eye/Face Wash Equipment. The standard sets similar requirements for combinations of emergency water delivery systems but states that "It is not necessary for all components to operate simultaneously (individual conditions will dictate this requirement)."

1.3.2.4 Supplemental Recommendations for Emergency Water Devices. We believe that the minimum flow rate for plumbed eyewash equipment should be at least 3 gallons per minute, and preferably 6–9 gallons per minute. (We believe that the ANSI minimum flow rate is suitable only for portable eyewash units for outside areas without a water supply.)

We believe that emergency conditions will often require *simultaneous* operation and use of both a shower and a drench hose, or a shower and eyewash unit, and that water supply and piping should be capable of providing for simultaneous operation of any combination installed.

We find that many OSHA Compliance Safety and Health Officers have been citing employers who have provided *only* plastic squeeze bottles for emergency flushing of chemicals from the eyes and face.

We recommend that emergency eyewash devices be located at sinks within the laboratory work area, but that emergency showers be located so that they will not be right in the middle of the chemical splash–spill area. If the laboratory space is subdivided into walled modules, we recommend that emergency showers be located in the corridors outside the modules. For the purpose of locating emergency showers, we recommend that the term "work area" be construed to include the corridors.

We believe that emergency eyewash devices should be located as close as possible to any area where chemicals may be handled and that location at a sink in the work area provides findability, convenient water supply and drainage, and easy testability at minimum cost of installation and dedication of space. We believe that laboratory personnel should be encouraged to use such emergency devices for any compatible purpose, such as using a drench hose to flush the sink routinely.

We believe that emergency showers can be located further from hazards than eyewash units or drench hoses. We recommend that emergency shower actuation be by a cord fastened to a wall so that the shower can be actuated by a person of any height and by someone in a wheelchair. We recommend that floor drains be installed close to each emergency shower and that consideration be given to automatic trap fillers or other plumbing arrangements so that traps in the floor drains do not dry out.

To facilitate effective flushing of chemicals from the body, we recommend that hand-held drench hoses be installed in combination with all emergency showers and that plumbing support simultaneous operation of both.

To address the question of water temperature for emergency showers, we recommend that the shower be installed on a cold water line to which there is a normally closed hot water line connection, with a valve which is sealed in the closed position with a breakable seal and a sign that indicates that the valve may be opened in case of emergency to temper the water.

1.3.3   Emergency Procedures

There are four emergency procedures for which all personnel should be trained: fire emergencies, clothing fires, spills, and chemical splashes.

All of these emergencies require action by whomever is in the vicinity and can respond immediately. Trained personnel or an emergency team can continue the emergency procedures after the initial action by personnel in the vicinity.

There are two problems that may complicate an emergency: (a) failure of personnel to respond promptly to the emergency and (b) failure of personnel to recognize the need to summon additional help.

Primary emergency procedures include the following steps:

1.   Alert personnel in the immediate vicinity of the emergency.
    a.   Give the nature and the extent of the emergency.
    b.   Give instructions:
        Call the local fire department.
        Sound the alarms.
        Close all doors.

2. Confine the emergency.

    a. Close doors to prevent spread of fire, smoke, vapor, gas, and fumes.

        Doors to corridor help to confine emergency to storeroom.

        Doors to stairwells help to confine emergency to one floor.

3. Evacuate the building or section involved.

    a. An evacuation alarm system is needed and generally required.

    b. Evacuation procedures should be posted.

    c. Assembly points should be designated for personnel accounting.

    d. Evacuation and assembly should be practiced in drills.

4. Summon assistance.

    a. Call the local fire department.

    b. Give the location and type of emergency.

1.3.3.1 Clothing Fire Emergency Procedures

1. Stop the person on fire from running! Do not allow anyone to run, not even to a fire blanket (Figure 1.12).

2. Drop the person to the floor or other horizontal surface. Standing will allow flames to spread upward. Standing in a fire blanket can funnel hot gases to the eyes and nose.

3. Roll the person to snuff out the flames. Blankets can be useful if they are brought to the person.

4. Cool the person. Remove smoldering clothing. Use water or ice packs to cool burns and minimize injury.

5. Summon medical assistance.

1.3.3.2 Spill Emergency Procedures.

The initial procedures for toxic chemical spills are the same primary procedures as those recommended for fires and other emergencies:

FIGURE 1.12   Fire blanket case.

Alert personnel in the vicinity

Confine the emergency by shutting doors

Evacuate the emergency area

Summon assistance

If the spilled chemical is as toxic and hazardous as chloroform, self-contained breathing apparatus (Figure 1.13) will be *essential* to protect

FIGURE 1.13 Essential equipment for a laboratory spill response team. Two sets of self-contained breathing apparatus, spare cylinders, personal protective equipment and clothing, and emergency rope. Complete sets of protective equipment are available in an area where spills will not prevent access to or donning of the equipment.

cleanup personnel from concentrations that can be immediately dangerous to life or health.

After cleanup personnel have been protected with necessary protective clothing and breathing apparatus, there are two approaches to dealing with the chemical spill: (a) use the floor as the reaction vessel for neutralization; and (b) absorb the chemical and carry out the reaction elsewhere.

Several of the spill kits that have been sold use the principle of neutralization on the floor. In some cases, spills may require neutralization in place, particularly if the material has splashed on the walls and the ceiling.

Having spill kits will provide a misleading sense of confidence if their capacity is less than that of the quantities that can be spilled. For example, many common spill kits have a rated capacity of only 4 ounces.

Having spill absorbent material in adequate amounts for possible spills is important, but the next problem is to have ways of distributing the material effectively over the spill, not just in isolated piles.

Bear in mind that when you pick up spilled hazardous material, you have just created hazardous waste that will have to be handled according to the regulations of the Resource Conservation and Recovery Act.

NFPA 43C requires special equipment and instruction if more than 100 lb (45.4 kg) of gaseous oxidizing material is stored. Respiratory protective equipment and other appropriate protective equipment must be readily available outside of the storage area, and training must be provided in methods for the safe handling of cylinders or other containers and for emergency procedures. Training is to include the use of respiratory and other personal protective equipment, drills on alarm and evacuation procedures, the handling of leaks, and plans for protection in case of fire in adjacent areas.

For storage of pesticides, the NFPA standard requires written procedures for preventing emergencies and for management of post-fire problems.

1.3.3.3   Chemical Splash Emergency Procedures.

The first step in any chemical splash is to get the concentrated chemical off the skin, and the second step is to desorb from the skin as much of the chemical as possible.

Reduction of chemical contact with the skin usually requires removal of contaminated clothing, to remove as much splashed chemical as possible and to prevent chemical on the clothing from being flushed through to the skin. Copious amounts of water are recommended and needed to dilute splashed chemicals.

Large amounts or pieces of water-reactive chemicals should be quickly brushed off the skin, if possible, but emergency water should be used for diluting and cooling so that such chemicals do not react with the moisture in the skin. We know of only one class of compounds that should not be flushed off with emergency water, the nitrogen mustards (chemical warfare compounds).

Desorbing chemicals from the skin will take approximately 10 times as long as the chemical was in contact with the skin. The generally recommended time for washing a splash is 15 minutes, which is a minimum. The time required for adequate flushing may be as long as several hours.

Immediate use of emergency water takes precedence over transporting the injured person to a medical facility.

Prevention of chemical splashes is the best procedure, by use of eye and face protection and protective clothing.

1.4 DISPENSING

Dispensing chemicals safely may require special equipment and procedures to prevent leakage, vapor dispersion, or fire. Leakage from dispensing containers or overfilled containers can cause damage or hazardous conditions in the dispensing area or in locations to which filled containers are taken. Dispersion of vapors can cause corrosion, adverse health effects or accumulation of flammable concentrations of vapors. Fire or explosion can be caused by ignition of flammable vapors by portable ignition sources, static electricity, or fixed electrical equipment.

Dispensing of chemicals may also require special equipment and procedures if personnel are working alone in dispensing areas. Emergency equipment and an alarm system should generally be provided in case of spill, splash, or fire.

Dispensing of flammable and combustible chemicals has special safety needs, many of which are required by codes developed by the National Fire Protection Association and adopted by governmental jurisdictions. For example, no flammable or combustible liquid should be stored or transferred from one vessel to another in any exit corridor or passageway leading to an exit. Class I and II liquids must not be dispensed in general storage areas unless the dispensing area is suitably cut off from other ordinary combustibles or liquid storage areas, and dispensing is not permitted in cut-off rooms or attached buildings larger than 1000 square feet in area.

Whenever possible there be no dispensing within a flammable liquid storage room, and dispensing and transfer operations should be separated from storage areas containing flammable and combustible liquids. If a fire does occur as a result of dispensing, there will be less fuel and less damage if the fire cannot spread to stored material. If there are severe space limitations and storage and dispensing cannot be separated, however, recommendations for dispensing should be followed in addition to the recommendations for storage.

1.4.1   Preventing Leakage

There are four important aspects of preventing leakage as a result of dispensing operations: (a) controlling pouring and gravity dispensing with appropriate valves, (b) limiting any pressure applied to dispensing containers, (c) providing expansion space within all containers that are filled, and (d) protecting all large glass bottles from breakage.

Dispensing of liquids can be accomplished by pouring, gravity flow, pressure, or pumping.

Pouring safely from a container depends on an individual's ability to lift and hold the container and to control the angle and rate of flow. It is generally safer to provide equipment that will hold the container and allow it to be turned for pouring and that will return the container to an upright position after pouring.

Dispensing by gravity flow by means of a valve in the bottom of the dispensing container is convenient. However, the valve should not leak and it must be self-closing if it is used for dispensing any combustible or flammable liquid. This requirement is based on preventing uncontrolled flow of the liquids if the valve is left unattended.

Liquids can be transferred by use of a siphon, but the procedure must be carried out so that there is no personal contamination in starting the siphon and no excess flow if the siphon is left unattended. Starting a siphon flowing by pressurizing the container with a hand-squeezed bulb will not generate enough pressure to deform or rupture the drum.

On the other hand, pressurizing a shipping container with air from a compressed air line is hazardous because such containers can easily be overpressurized so that the air pressure ruptures them and blows liquids out of the container. Use of air pressure is prohibited for transferring any flammable or combustible liquid.

Transfer of liquids by pressure of inert gas is permitted by NFPA 30 only if the design pressure of the container is known and if controls and pressure-relief devices are provided to limit the pressure so that it cannot exceed the design pressure.

Air pressure can safely be used for transferring liquids if the air pressure operates a pump without pressurizing the container.

Pumping liquids from an opening in the top of the dispensing container avoids the necessity of tipping the container or positioning it horizontally for gravity dispensing and results in safer handling of drums and use of less floor space. Pumping will avoid leakage through a drain valve. One disadvantage is that a pump will be required for each liquid that could be contaminated by dispensing with a pump that is also used for other liquids.

Providing expansion space within all containers that are filled is important because overfilling containers can result in pressures sufficient to cause leakage or to rupture the container. Safety cans, which have a spring-loaded lid, will vent vapors if they are filled at a temperature less than that in the area to which they are taken and stored. Glass bottles with screw cap lids can rupture if they are filled nearly to the top with cold liquid and then stored in a warm or hot area.

Protective measures should be taken to prevent breakage of large glass bottles during storage or transportation within the facility. Plastic-coated bottles can be purchased that are suitable for transporting and using liquids that are heavy, corrosive, extremely flammable, extremely toxic, or very expensive. Two-piece plastic jackets are available for protecting some types of gallon reagent bottles. Several types of bottle carriers are also available.

Protection of large glass bottles of chemicals with break-resistant carriers or coatings can provide effective separation for containers of incompatible chemicals that could react if mixed.

1.4.2 Preventing Vapor Dispersion

Ventilation is needed for dispensing operations that may disperse vapors or aerosols that are corrosive, irritating, toxic, or flammable.

Corrosive vapors or aerosols can damage the storage facility, equipment, and other containers. Corrosive, irritating, or toxic vapors can cause discomfort or adverse health effects. Flammable vapors can accumulate in concentrations that, if ignited, could result in a flash fire or explosion.

The ventilation required by the NFPA Flammable and Combustible Liquid Code is primarily dilution ventilation to prevent accumulation of flammable vapors from leaking storage containers. Ventilation of flammable liquid storage rooms is required for floor areas where flammable vapors can collect.

The ventilation required by the NFPA Code is not designed to protect personnel from exposures in dispensing operations. Effective ventilation of dispensing operations will exhaust vapors from the point at which the vapors are dispersed, and away from the breathing zone of the person doing the dispensing (Figure 1.14).

The NFPA Code recognizes that local or spot ventilation may be needed for control of health hazards and allows such ventilation to provide up to 75% of the ventilation required by the NFPA Code for the room. The code requires one cubic foot per minute per square foot of floor area (0.028 m^3 per 0.0929 m^2).

The mechanical ventilation system for dispensing areas must, according to the NFPA Code, be equipped with an airflow switch that will sound an audible alarm if the ventilation fails.

The NFPA Fire Protection Standard for Laboratories Using Chemicals, NFPA 45-1982, specifies that transfer of Class I liquids to smaller containers from bulk stock containers not exceeding 5 gallons (18.9 liters) in capacity inside a laboratory building or laboratory work area shall be

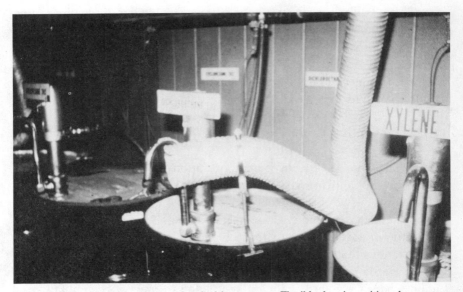

FIGURE 1.14 Dispensing solvents by air-driven pumps. Flexible duct is positioned to capture and exhaust vapors released during dispensing.

made: (a) in a laboratory hood, (b) in an area provided with ventilation adequate to prevent accumulations of flammable vapor–air mixtures exceeding 25% of the lower flammable limit, or (c) in a separate inside storage area, as described in NFPA 30.

Transfer of Class I liquids from containers of 5 gallons (18.9 liters) or more capacity must be carried out in a separate area outside the building or in a separate area inside storage area that meets NFPA requirements.

Standard NFPA 45 explains that ventilation for transfer operations should be provided to prevent overexposure of personnel transferring flammable liquids and that control of solvent vapors is most effective if local exhaust ventilation is provided at or close to the point of transfer.

Explosion venting is not required for separate inside storage areas if containers are no greater than 60 gallons (227 liters) and if transfer from containers larger than 1 gallon (3.785 liters) is by means of approved pumps or other devices drawing through a top opening, according to NFPA 45. 'For gaseous oxidizing materials, positive exhaust ventilation shall be provided for an enclosed storage area. (Natural ventilation is adequate for open storage areas.)'

1.4.3   Preventing Ignition of Vapors

The NFPA Flammable and Combustible Liquid Code requires that precautions be taken to prevent the ignition of flammable vapors and states specific requirements for controlling ignition sources where flammable or combustible liquids are dispensed. Ignition sources include open flames, smoking material, cutting and welding operations, hot surfaces, radiant heat, frictional heat, static electricity, electrical and mechanical sparks, spontaneous combustion, and heat-producing chemical reactions.

Hot work, such as welding or cutting operations, use of spark-producing power tools and chipping operations should not be permitted except under supervision of a responsible individual who will make an inspection of the area before work begins to be sure that it is safe for the work to be done and that safety procedures will be followed.

Static electricity is generated when liquids are dispensed, and under some conditions it may accumulate to voltages high enough to cause discharges that can ignite flammable vapors. Class I liquids, and Class II or Class III liquids at temperatures above their flash points, must not be dispensed from a metal container into a metal container unless there is an electrical connection between the containers, by maintaining metallic contact during filling or by a bonding wire between them or by other conductive path with an electrical resistance not greater than 10 ohms. Such electrical connection or bonding is not required where a container is filled through a closed system or one of the containers is made of glass or other nonconductive materials.

Electrical wiring and equipment located in inside rooms used for storage of Class I liquids must be suitable for Class I, Division 2 classified locations. Such equipment is commonly called explosionproof (Figure 1.15).

FIGURE 1.15   Solvent dispensing room with a grounding system, ventilation, automatic fire extinguishing system, explosionproof lights, and smoke venting.

In rooms where dispensing of Class I liquids is permitted, electrical systems must be suitable for Class I, Division 2 classified locations, except that within 3 feet (0.914 meters) of a dispensing nozzle area, the electrical system must be suitable for Class I, Division 1 locations.

The NFPA Code specifies that electrical equipment in ventilated areas used for drum or container filling be labeled by the manufacturer as suitable for Class I locations, according to the National Electrical Code. The NFPA Codes specify that the electrical equipment within 3 feet in all directions from filling and venting openings must be suitable for Division 1 locations, and that the equipment located 3–5 feet from vent or fill openings must be suitable for Division 2 locations. Electrical equipment up to 18 inches above floor or grade level within a horizontal radius of 20 feet from vent or fill openings must also be suitable for Division 2 locations.

1.4.4 Working Alone

Since working alone in a dispensing area can be hazardous in case of accident or overexposure, extra precautions are usually needed. Precautions should be related to factors such as the volume of material dispensed during a given period of time, the hazards of the materials being dispensed, the size of the dispensing area and whether a person working in the dispensing area can be seen by someone outside the area.

Extra precautions do not seem necessary if the dispensing consists solely of someone going into the area occasionally to fill a small container such as a liter bottle with a material with no unusual hazards.

Extra precautions do seem necessary if filling is done frequently, if one person works for a long time in the area, or if the dispensing area is located remotely or in an area where there is infrequent pedestrian traffic.

Where the interior of a dispensing room cannot be seen from outside the room, it is a common precaution to keep the door open whenever anyone is inside the room.

If it is not feasible to have a second person present during dispensing operations, the hazards may warrant an audible monitoring system that will allow another person to hear immediately any loud noise or call for help.

Employees should not work alone when connecting or disconnecting cylinders or other containers of gaseous oxidizing materials.

1.4.5   Emergency Alarms

In addition to the emergency equipment recommended for controlling chemical splashes and spills, each dispensing area should be provided with an emergency alarm station outside the dispensing area. The emergency alarm station could consist of a conventional fire alarm pull station, a special emergency pull station, or an emergency telephone that sounds an alarm and is answered when the telephone is picked up.

1.5   INVENTORY

1.5.1   Inspection of Storage Containers and Facilities

Management of chemical storage should include periodic inspection of storage containers, shelving, and storage facilities. Inspection should determine whether there has been any corrosion, deterioration, or damage as a result of leakage or spills from containers. Containers should also be inspected to ensure that labels are legible and fastened to their containers.

Information has been published on the hazards of prolonged storage of peroxidizable compounds, picric acid, and some other materials that become dangerous or deteriorate over a period of time. However, there is no published information on the length of time that chemicals in unopened containers can safely be stored. How long chemicals can be safely stored unopened will depend partly on the chemical and partly on the environmental conditions in the storage area.

Chemicals that lose their identification because their labels deteriorate or fall off cannot safely be used or disposed of without analysis to determine their identity and hazards (Figure 1.16).

1.5.2   Dating of Laboratory Reagents

Management of chemical storage should include dating of all containers when received and when opened. Dating of containers of laboratory reagents is usually required so that materials that have a limited shelf life can be replaced to avoid interference with accurate test results. Equally important is the dating and management of materials that can deteriorate and produce severe hazards. For example, picric acid can dry and ethers

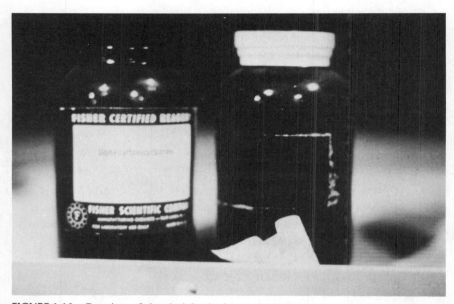

FIGURE 1.16   Container of chemical that is about to become an unknown and a costly waste.

can peroxidize, creating explosive hazards. It is critical that each container of material that can deteriorate in prolonged storage be dated when it is received for storage and that peroxidizable materials be dated when they are first opened.

Ethers and other peroxidizable compounds kept for prolonged periods after they have been opened will form peroxides that can react explosively when the container cap is removed or when they are concentrated during laboratory activities. In one case, the person removing the stopper of an old bottle of ether was killed when the peroxides in the closure were detonated by the friction of opening the container.

Ethers should be managed safely by the following procedures:*

1. Buy ethers in the smallest size possible (to limit amounts exposed to air).

*Recommendations for identifying peroxidizable compounds, limiting shelf life, testing for peroxides, and removing peroxides are available in an article entitled "Control of Peroxidizable Compounds," appearing in Volume 3 of *Safety in the Chemical Laboratory,* published by the Division of Chemical Education, Publications Coordinator, 215 Kent Road, Springfield, Pennsylvania 19064.

2. Date all ethers when they are opened.

3. Test all opened ethers for peroxide concentration within a few months from the date the ethers are opened and at regular intervals thereafter. Test ethyl ether and dioxane every 6 months during storage and test isopropyl ether every 3 months.

4. If peroxide concentrations are acceptable, redate the container and retest at the next scheduled test date.

5. If peroxide concentrations are not acceptable, remove the peroxides or dispose of the ether as hazardous ignitable waste.

6. If peroxide concentrations cannot safely be determined but it is believed they may be excessive and possibly explosive, plan to have the entire container removed with extreme care as a shock-sensitive explosive hazard.

1.5.3   Disposal of Shock-Sensitive Chemicals

If you find shock-sensitive compounds that need to be disposed of, be sure that the disposal team recognizes the hazards and plans the disposal in ways that will not threaten personnel or facilities. It is advisable to schedule the removal of potentially explosive compounds when it will be possible to evacuate the area or when there will be a minimum exposure of personnel (such as late at night).

Ethers with high concentrations of peroxides are extremely sensitive to impact and physical shock and are capable of violent explosion. Ethers with unknown concentrations of peroxides should be handled as if they were booby-trapped. Avoid letting anyone "shake" the container to see how much is in it. No shock-sensitive compound such as old ethers or old picric acid should be handled with disregard for their explosive power.

1.5.4   Management of Space

When a chemical storage space needs to be used to the maximum, compatible materials should be stored by the size of the container. For example, a considerable amount of potential storage volume is wasted if shelves spaced vertically for storage of gallon bottles are used for storage of small bottles.

The ability to select the appropriate storage location and to find and manage stored chemicals will be greatly improved by establishing a system for identifying and coding both the chemicals and the storage spaces. If laboratory chemicals and their containers can be identified by a unique number (such as the *Chemical Abstracts* registry number), and if the storage locations can be identified by numbers, the chemicals can be stored in a safe and retrievable manner with less confusion and with less possibility of losing track of the chemicals in an alphabetical storage arrangement.

For example, one company with a large research organization and many organic chemicals has set up a system in which:

On the top of each small bottle is a unique code number, in addition to the chemical name on the side of the bottle.

Small bottles are stored in drawers in a ventilated cabinet.

Each cabinet, drawer, and section has an identifying letter or number.

Each chemical is locatable by reference first to an alphabetical list that is computer-generated and cross-indexed with commonly used terminology. The reference list gives the unique chemical code number and the coordinates of the storage location. The chemical can be obtained by going then to the storage room, cabinet, drawer, and drawer section. The code number on top of the bottle identifies the chemical, and the name on the bottle confirms that the search has been successful.

CHAPTER 2

INCOMPATIBLE CHEMICALS IN THE STOREROOM: IDENTIFICATION AND SEGREGATION

LESLIE BRETHERICK,
Chemical Safety Matters, Berkshire, England

2.1   INTRODUCTION

2.1.1   Previous Developments

The need to exercise control over the physical arrangement of chemicals in storerooms so as to minimize the consequences of accidental mixing of incompatibles by spillage or breakage, or in storeroom fire, has been recognized for a long time. One of the earliest published references to this need was that of Davison (1), who in 1925 proposed that safety considerations should be included when planning and implementing educational chemical storage facilities. In 1951 the Los Angeles Fire Department composed a Dangerous Chemicals Code, which still forms the basis of the list for segregating incompatible chemicals in the recent ACS booklet (2). An extended list of incompatible chemicals by Fawcett (3) has been reproduced extensively since it appeared in 1952.

Voegelein (4) gave a considerable amount of detail on the containers and storage conditions appropriate to a fairly wide range of dangerous chemicals, but only very general guidance on their segregation in storage. The most recently published guide, that by Pipitone and Hedberg (5), covers in detail many aspects of safe storage of chemicals on the relatively small scale. This reference again emphasizes the point made 18 years previously by Steere (6) that many potential problems may arise from simply arranging containers of chemicals in alphabetical order and ignoring incompatibilities.

The advent of new technology holds promise in providing additional data on the hazard potential of chemicals in storage. Using a computer program entitled CHETAH, Coffee (7) and Treweek (8) have developed mathematical procedures for predicting the self-explosive potential of organic chemicals and the energy release from binary systems of incompatible chemicals.

In spite of these reports and much more good advice that has been available for so long, however, the results of recent surveys (5,9) show that any reasonable level of segregation in storage of chemicals appears rather exceptional among many professional chemists and educators.

2.1.2   The Need for a Positive Simple Basis

While pondering on this obviously undesirable state of affairs, two factors emerged as possible major contributors: (a) the preponderance of the almost universally negative aspects of advice available on chemical segregation—there is unlimited advice on what chemicals cannot be stored together, but infinitesimal information on those chemicals that can be stored together in relative safety; and (b) the considerable degree of complication and uncertainty that surrounds the question of how best to classify chemicals to allow a suitable system of segregation in storage to be developed.

There seems to be no clear consensus on what and how many classes or groups of chemicals exist that need to be segregated. Ten commonly mentioned incompatible groups are flammables, oxidants, reducers, concentrated acids and bases, water-reactives, toxics, peroxidizables, pyrophorics, and cylinder gases. The incompatibilities in the first five groups arise from the potential for exothermic, violent, or even explosive reactions on accidental intermixing. The presence of the sixth group of water-reactive compounds during water-based fire-fighting operations could lead to severe complications. Toxic materials often need physical control on their distribution for use, and for those of high volatility, special ventilation may be required. Peroxidizable materials need cool, dark, and (usually) air-exclusive storage, whereas pyrophorics effectively have a built-in ignition system, needing only contact with air (or sometimes water) to establish the triangle of fire. The tenth group requiring segregation—the cylinder gases—is exceptional in that as well as the hazard inherent in the contents of a particular cylinder, there is often a high kinetic energy content due to the state of compression of the contained gas.

However, these 10 groups are not mutually exclusive. Table 2.1 gives some examples of chemicals that belong simultaneously to at least two of these groups and that would, therefore, be difficult to classify for segregation solely on that basis. An additional assessment of the major hazard would be necessary.

More elaborate classification systems were proposed for classifying chemicals for segregation in storage, especially on the large scale. The J. T. Baker chemical safety course, based on an academic study (10), suggests 16 classes with considerable emphasis on inorganic–organic differentiation. The U.S. Coast Guard, considering the rather special requirements for safe shipboard and often bulk transportation and storage, uses no fewer than 43 classes (with several additional restrictions) as the basis of the CHRIS proposals for complete segregation (11)

Such a high degree of segregation is clearly impracticable and largely irrelevant to the needs of a relatively moderate-sized storage facility nec-

TABLE 2.1 Group Classification Problems of Chemicals[a]

Groups	Flammable	Acid	Base	Oxidant	Reducer	Gas	Toxic	Peroxidizable	Pyrophoric
Water-reactive	Acetyl chloride	Chlorosulfuric	Potassium hydroxide	Chromyl chloride	Lithium aluminum hydride	Boron trichloride	Phosgene	Acryloyl chloride	Trimethyl-aluminum
Flammable		Thio-acetic	Methyl-amine	Calcium hypochlorite	Calcium hydride	Butane	Hydrogen cyanide	Tetrahydro-furan	All[b]
Acid			Betaine	Nitric acid	Formic acid	Hydrogen chloride	Hydrofluoric acid	Acrylic acid	
Base					Hydrazine	Ammonia	Dimethyl-amine		Trisilyl-amine
Oxidant					Redox salts	Fluorine	Bromine		
Reducer						Hydrogen	Hydrazine		Titanium hydride
Gas							Arsine	Vinylidene chloride	Diborane
Toxic								Acrylo-nitrile	Phosphine
Peroxidizable									All[b]

[a]Difficulties in classifying hazardous chemicals using the usual suggested groups are listed in columns 2 through 9. Each of the chemicals shown belongs to at least two groups simultaneously.

[b]Note that all pyrophorics are flammable and peroxidizable, but the converse is not true.

50

essary to serve a small to medium group of laboratories engaged on relatively small-scale research or teaching activities.

2.2 FLAMMABILITY AND WATER COMPATIBILITY AS MAJOR CRITERIA

2.2.1 Flammability

Of the total number of accidents involving the storage of chemicals, the most serious fraction involves fire as the initial acute hazard, usually causing considerable financial loss and occasionally injury or fatality. The possibility of fire and the implications of measures for effective prevention or control should play, therefore, a major part in the overall strategy of segregating chemicals in storage. This is not a new concept (12), but it does not seem to have been applied widely other than to stores (often bulk stores) for highly flammable organic solvents.

2.2.2 Water Compatibility

Water is the most appropriate and effective extinguishing medium for some types of fire involving chemicals, but the least appropriate to other types, so compatibility with water emerges as a further important criterion for storage considerations. Those combustible materials, liquid or solid, that are soluble in or denser than cold water are considered to be *water compatible*. Those combustible materials, liquid or solid, that are insoluble in and are less dense than (i.e., float on) cold water are considered as *water incompatible*, as are those incombustible chemicals that react adversely with water. Special fire extinguishers (13) are necessary for some combustibles incompatible with water such as metallic sodium or calcium hydride.

Although suppliers' catalogs or the labels on currently supplied containers of chemicals will usually indicate flammability, toxicity, and sometimes the density of the contents, it may be necessary to obtain water solubility and reactivity data from a literature source (14–20). Thus a decision on compatibility or incompatibility with water can be made properly with any further subclassification aspects necessary for the scheme outlined below.

2.3   IDENTIFICATION AND SEGREGATION OF HAZARDS

2.3.1   Identification and Classification

In view of its importance in overall storage considerations, flammability has been adopted as the major basis of division of the hazard classification proposed in this chapter. For reasons outlined previously, compatibility with water has been adopted as the next most important basis. Taken in combination, four major areas are produced:

Area 1:   Flammables that are compatible with water.

Area 2:   Flammables that are *incompatible* with water.

Area 3:   *Non*flammables that are compatible with water.

Area 4:   *Non*flammables that are *incompatible* with water.

Because the risk of fire in flammable materials is closely related to the volatility (as defined by the flash point), the criterion of flammability for the present purpose has not been set on the absolute basis of combustibility or noncombustibility, but rather on a relative basis. Thus those materials (usually organic liquids) with flash points below 37.8°C (100°F, Class IA, B, or C, labeled "Flammables") in the United States, or below 55°C (131°F, labeled "Extremely Flammable," "Highly Flammable," or "Flammable") in Europe, are included within the "Flammable" grouping of Areas 1 and 2 for the present purpose. Materials with flash points above those values are included as "Combustibles" of relatively low fire hazard, which may be stored with noncombustible materials of the same water-compatibility type in Areas 3 and 4.

Some degree of further physical segregation is required within Areas 1–4 for those toxic or reactive materials that need more closely controlled storage conditions than normal containers on open shelves. This aspect is discussed later.

There are four other groups that require separated and specific storage conditions. Three of these groups have major potential hazards, although limited quantities are usually kept in storage. These are:

Area 5:   Materials that become unstable above ambient temperatures.

Area 6:   Materials unstable (or too volatile) at ambient temperature and which require refrigerated storage.

Area 7:   Pyrophoric materials for which loss of containment will cause ignition (see Section 2.1.2).

Area 8:   Compressed gases in cylinders (special hazard; see Section 2.1.2).

Some degree of further chemical segregation usually will be necessary within Areas 5, 6, and 8. If a moderate amount of high-hazard material is necessary for the work in hand, or if many smaller quantities of different materials of similarly high hazard are to be stored, those chemicals may need storage in several separate cupboards or enclosures within the appropriate area.

2.3.2   Main Levels of Segregation

The eight main groups identified above are shown in schematic form as storage areas in Figure 2.1. With some secondary subdivision and separation, the author proposes that those main groups form the basis of an effective segregation plan for a wide range of chemicals in up to kilogram quantities. These amounts are expected to be necessary for a group of laboratories engaged in normal small-scale chemical research or teaching activities. Specialized laboratories requiring larger quantities or proportions of particularly hazardous materials would need to adapt the general scheme shown in Figure 2.1 by accentuating the appropriate area, even to the extent of complete separation to accommodate the greater chemical volume.

It is emphasized that the schematic layout of Figure 2.1 is intended only to illustrate the general principles and is *not* a ground plan for a practical layout (see Section 2.4.2). Relatively cool temperate conditions, free from direct sun exposure and excessive cold in winter, are assumed to prevail in all areas except Area 6.

If the flameproofed Areas 1 and 2 are employed to accommodate bulk amounts of flammable solvents (i.e., in large quantities not easily handled as laboratory packages as in the case of 55-gallon drums), storage and dispensing of those solvents should be separated from the main storage areas, minimally by fire-resistant partitions. Those bulk areas must be equipped with drum grounding facilities and accessed by means of separate wide doors for adequate drum handling. Such bulk areas are outlined schematically in Figure 2.1.

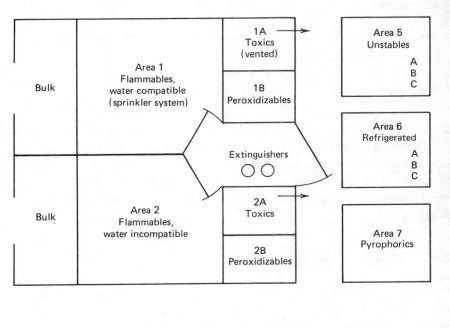

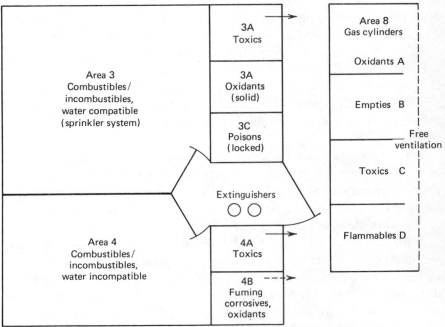

FIGURE 2.1 Outline of schematic arrangements for segregated storage of a wide range of laboratory chemicals. (*Note:* This is not intended as a working ground plan.)

The relatively large size of Area 3 reflects the probability that the great majority of relatively nonhazardous solid (and particularly inorganic) materials would tend to be stored in a water-sprinklered area.

2.3.3 Sublevels of Segregation Within a Storage Area

What is to be done with those chemicals that belong to the same area but are incompatible with each other? The author has adopted a practical point to simplify some of the segregation that should occur *within* each area. There is a relatively low probability that a solid, spilled or released through breakage, will effectively commingle with another incompatible solid because most solids do not flow readily. On the other hand, there is a higher probability that two incompatible liquids, spilled into accidental contact, will mix effectively with deleterious results.

Therefore, there could be overall advantage in storing solids separately from, and on shelves above, liquid chemicals within the same compatibility class. Furthermore, the priority to segregate pairs of incompatible liquids, and solid–liquid incompatible combinations, is higher than to segregate pairs of solids that may not be compatible.

The enclosed fire-resistant storage areas for toxic materials (Areas 1A, 2A, 3A, and 4A) shown vented to outside would need a modest ventilation rate. Dispensing of toxic materials from the storage containers is left to be done in well-ventilated laboratory hoods.

Deciding whether a material is of sufficient toxicity to merit removal from an open-shelf location and placement into the enclosed toxic storage area could be based on the vapor hazard index (VHI) recently proposed by Pitt (21). The index is a measure of the factor by which the concentration of a substance in a vapor-saturated atmosphere at a particular temperature exceeds the threshold limit value (TLV) or permissible exposure limit (PEL). The temperature in this application would be a typical ambient value, 20 or 25°C. Use of the VHI rather than TLV or PEL as the decision basis for segregation would then include an indication of the risk of inhalation likely to be present in a spill or breakage incident in storage. The VHI can, of course, be calculated only for those several hundred common materials for which TLV or PEL figures are available.

Although the enclosed, cool and dark Areas 1B and 2B for storage of flammable peroxidizable compounds are schematically shown the same size as the adjacent toxic storage enclosures seen in Figure 2.1, the actual sizes are determined by the ratio of contents necessary for the particular laboratory or laboratories using the storage area. If the absolute quantities of toxic materials and peroxidizables were very small, a feasible arrangement would be to use an unvented dark lower part of a single cupboard for peroxidizables, with the toxics located in the separate, vented top section.

The separate locked cupboard representing Area 3C (for poisons) is proposed as a specific provision for those highly toxic or carcinogenic materials requiring mandatory control on access and use. In some cases where nonvolatile poisons are used, it may be more appropriate for Area 3C to be located in a laboratory under senior supervision, so that necessary registration formalities can be effected.

An interesting point emerged during consideration of the classification and segregation of oxidizers, which previously have tended to be lumped together as one internally compatible group. There is, in fact, need to exercise considerable care, and there seems a good case, not only for segregating water-compatible oxidizers (Area 3B) from water-incompatible oxidizers (Area 4B), but also solid oxidizers from liquid oxidizers within both these groups on the grounds of potential reactivity.

Area 4B is devoted to the storage of materials (usually chemicals containing halogen and including some oxidizers) that react with moisture to produce toxic fumes that are also corrosive to structural materials. Ideally, the primary containers of such materials should not only be sealed with a reagentproof closure, but also contained (not necessarily singly) within a secondary desiccated enclosure to minimize deterioration of the contents and concomitant fume generation. A proprietary "static hood" (22) is available for this purpose, but snap-lid plastic food containers with silica gel and some granular active carbon will serve equally well, particularly if the lid is also sealed round with PVC (not cellulose) adhesive tape. If such storage measures are used in Area 4B, the ventilation via negative pressure hinted at by the dotted arrow in Figure 2.1 should be unnecessary. Spent desiccant and carbon from the secondary storage boxes should be discarded rather than regenerated by heating. If total quantities are small, Areas 4A and 4B might be consolidated, using shelf segregation, into a single ventilated enclosure.

Secondary Areas 5–8. For materials stored in Area 5 (those unstable at elevated temperatures), the main requirement is isolation from heat sources. A location remote from fire-risk areas is essential. This is also true of Area 8. Adjacent placing of those two areas might be feasible.

The refrigerated Area 6, for storage of materials unstable at ambient temperature, or highly volatile toxics, will inevitably be a refrigerator or cold room, with two essential features: (a) a spark-free interior (external thermostat contacts) and (b) an individual power supply to minimize the risk of power failure and subsequent warming. Refrigerated storage is often provided in laboratories for convenience in supervision, but if located remotely, some device to warn personnel of power failure is necessary so that remedial action can be taken. Minimal internal segregation of materials in the refrigerated store would require separate shelf trays (with edge spaces for cold circulation), perhaps with additional secondary containment for volatile corrosives as in Area 4B. Additional refrigerators will be necessary for storage of more than minimal quantities of highly incompatible thermally unstable materials.

Since Area 7 (for storing pyrophoric materials) has the potential for high fire risk, separation by a safe distance from flammable or heat-sensitive materials is of paramount importance. A considerable degree of secondary containment, preferably under dry inert atmosphere (nitrogen-filled desiccators), will be necessary to prolong the storage life of most highly reactive materials of this group. Subsidiary storage of pyrophorics in laboratories is also likely to occur, because small (working) quantities of pyrophoric reagents and catalysts are usually maintained for safety and convenience in use in the inerted glove boxes used in laboratories engaged in organometallic chemistry. It seems essential to exclude from Area 7 those few pyrophoric materials (white phosphorus, zirconium powder) that need water cover for safe storage, in view of the water incompatibility of most pyrophorics. No dispensing of pyrophorics is permissible in storage Area 7.

Isolation from potential fire sources is the main requirement for cylinder storage in Area 8, to minimize the possibility of cylinder rupture and release of the often highly compressed gaseous contents. A secondary requirement is that the whole area must be freely ventilated to permit the dispersion of any toxic or flammable gas that might leak slowly through a faulty cylinder valve, which is a not too uncommon occurrence in practice. Segregation of cylinders into the four indicated sections is desirable if more than a dozen or so cylinders are stored. The specific area for "empties" (which should invariably contain a small residual pressure to minimize the possibility of air contamination of the cylinder) is an important feature of the arrangement. The securing of cylinders by retaining straps or chains is equally important in storage as it is in the use of cylinders. Cylinders of inert gases may be dispersed as convenient into the other full cylinder storage sections.

2.4 PRACTICAL APPLICATIONS

2.4.1 Classification Applied to a Range of Chemicals

To check the general validity of the proposal developed above and to provide examples, particularly of the sublevels of segregation that may be necessary within each area, a selection of some 200 chemicals that might be expected to be found in typical storerooms serving multiproject research laboratories has been classified into the eight major areas discussed. Main levels of segregation within each area, and any points of detail related to these, are indicated in the groupings below by the following abbreviations:

o = Oxidizer   n = Not otherwise classified
r = Reducer   p = Peroxidizable
a = Acid   d = Store dark
b = Base   Note that each prefix also applies to all unprefixed
                 items listed below it.

Specific incompatibilities are indicated by an abbreviation in parentheses following the name of the chemical:

Area 1—Flammables (Compatible with Water)

Solids	Liquids	
o Ammonium dichromate	o Nitromethane (no b)	Acetonitrile
Sulfur powder	r Hydrazine hydrate (b)	Chlorobenzene
	a Acetic acid	Ethanol
	Phosphinic acid (r)	Methanol
	b t-Butylamine	Methyl isobutyl
	Pyridine	ketone
	n Acetone	Propanol

Area 1   (*Continued*)

1A—Toxics

Solid

Acrylamide (d)

Liquids

Acetone cyanohydrin (no a,b)
Acrylaldehyde (p)
Allyl alcohol
Allylamine
Allyl chloride
Allyl chloroformate
Carbon disulfide
Epichlorohydrin

1B—Peroxidizable

Crotonaldehyde
1,1-Dimethoxyethane
1,2-Dimethoxyethane
Dioxane
Ethyl acrylate
Tetrahydrofuran

Area 2—Flammables (Incompatible with Water)

Solids	Liquids
r Aluminum powder	Acetic anhydride
Calcium hydride	Acetylacetone
Lithium aluminum hydride	Amyl acetate
Magnesium powder or turnings	*t*-Butyl chloride
Sodium hydride	Cyclohexane
Sodium dithionite	Ethyl acetate
n Phosphorus pentasulfide	Petroleum ether
	Toluene

2A–Toxics

Acetyl chloride (store dry)
Acrylonitrile (p)
Benzene
Toluene-2,4-diisocyanate

2B—Peroxidizables

Solid	Liquids
Potassium (no CO_2)	Dibutyl ether
	Diethyl ether
	Styrene
	Vinyl acetate
	Vinylidene chloride

59

Area 3—Combustibles and Incombustibles (Compatible with Water)

Solids

r Hydrazine sulfate
  Hydroxylamine hydrochloride
a Adipic acid
  Benzoic acid
  Chloroacetic acid
  Citric acid
  Cyanoacetic acid
  Maleic anhydride (no b)
  Oxalic acid
  Phenol
  Sulfamic acid
  p-Toluenesulfonic acid
b 2-Aminopyridine
  Calcium oxide
n Aluminum oxide
  Aluminum sulfate
  o-Aminophenol

Ammonium chloride
Ammonium fluoride
Ammonium sulfate
Ammonium thiocyanate (d)
Anthracene
Biphenyl
Calcium carbonate
Caprolactam
1-Hydroxy-2,2-bisphenylpropane
p-Chlorophenol
Cobalt(II) acetate
Copper sulfate
p-Nitrotoluene
Sodium azide (no a)
Sodium sulfide (no a)
Zinc powder
Zinc chloride

Liquids

o Sodium hypochlorite (no a)
  Nitric acid (a)
  Perchloric acid (a)
r Formaldehyde solution (no HCl)
  Formic acid (a)
a Hydrochloric acid (no o)
  Phosphoric acid
b Ammonia solutions
  Butylamine
  Cyclohexylamine
  Ethanolamine

  Morpholine
b Sodium hydroxide solutions
n Acetophenone
  Aniline
  Butoxyethanol
  Carbon tetrachloride
  Chloroform
  Dibutyl phthalate
  Furfural
  Nitrobenzene
  1,1,1-Trichloroethylene

Area 3 (*Continued*)

3A—Toxics

Solids

2,4-Dinitrochlorobenzene

Liquids

Acrylic acid (p)	Bromine
Adiponitrile	Chloropicrin
Benzenethiol (stench)	Dimethylformamide
Benzonitrile	Dimethyl sulfate
Benzoyl chloride	Hexamethylphosphoramide

3B—Solid Oxidizers (Liquids in Area 3)

Ammonium nitrate	Manganese dioxide
Ammonium perchlorate	Potassium chlorate
Ammonium persulfate	Potassium dichromate
Barium peroxide	Potassium nitrate
Calcium hypochlorite	Potassium permanganate
Chromium trioxide	Silver nitrate (d)
Dichloroisocyanuric acid	Sodium nitrate
Lead nitrate	Uranyl nitrate
Magnesium perchlorate	

3C—Poisons (Including Carcinogens, in Locked Cupboard)

Arsenic trioxide	Cadmium oxide
Barium chloride	β-Naphthylamine
Benzidine	Potassium cyanide
Beryllium oxide	Thallium nitrate

Area 4—Combustibles and Incombustibles (Incompatible with Water)

Solids

o Sodium peroxide	o Sulfuric acid	Cyclohexane
r Calcium	n Anisole	Decahydronaphthalene
Sodium borohydride	Benzaldehyde	Dimethylaniline
b Potassium hydroxide	Benzyl alcohol	Toluidine
	Butanol	Xylidine
	Cumene	

4A—Toxics

Solid	Liquid
γ-Hexachlorocyclohexane	Dimethylcarbamoyl chloride

4B—Fuming Corrosives (Store Dry)

Solids	Liquids	
Aluminum chloride	Acetyl bromide	Methanesulfonyl chloride
Antimony trichloride	Antimony	Oleum
Cyanogen bromide	pentachloride	Phosphorus tribromide
p-Toluenesulfonyl	Benzotrichloride	Phosphoryl chloride
chloride	Benzoyl chloride	Silicon tetrachloride
	Chlorosulfuric acid	Sulfinyl chloride
	Chromyl chloride (o)	Titanium tetrachloride
	Ethyltrichlorosilane	

Area 5—Materials Unstable Above Ambient Temperature

5A—Oxidizers

Cumene hydroperoxide
Hydrogen peroxide (vent container)
Peroxyacetic acid

5B—Organics

m-Dinitrobenzene
Picric acid (store wet)

Area 6—Refrigerated Materials Unstable or Too Volatile at Ambient Temperature

6A—Oxidizers

Dibenzoyl peroxide
Di-*t*-butyl peroxydicarbonate

6B—Toxics	*6C—Flammable*
Methyl fluorosulfate	Acetaldehyde
Methyl iodide	Dimethyl ether

Area 7—Pyrophorics

Butyllithium solutions
Diethylzinc
Triethylaluminum solutions

Area 8—Cylinders of Compressed or Liquefied Gases

8A—Oxidizers	*8C—Toxics*	*8D—Flammables*
Chlorine	Arsine	Acetylene
Oxygen	Boron trichloride	Ethylene oxide
Ozone solutions	Phosphine	Hydrogen
	Sulfur dioxide	

8B—"Empties" (Slight residual pressure)

2.4.2 Further Practical Necessities

As has been stressed previously, the general arrangement presented in Figure 2.1 and discussed in some detail above deals with the development of a general principle for the segregation of chemicals in small-scale storage.

To apply this principle practically to an existing storage facility or to design a new one will, especially in the latter case, require much detailed consultation and consideration of many specific factors including local geography, codes, and fire regulations. Some of the major practical factors are as follows, with pertinent references given in parentheses: planning and design of chemical storage facilities (23, 24), segregation of chemicals (2, 3, 16–19, 25), fire protection in chemical storage (26), chemical container and labeling requirements (16, 27), record keeping (5), and storage of gas cylinders (28, 29).

REFERENCES

1. H. F. Davison, *J. Chem. Educ.*, **2**, 782 (1925).
2. ACS Committee on Chemical Safety, *Safety in Academic Chemical Laboratories*, 3rd ed., ACS, Washington, DC, 1979, Appendix 3.
3. H. H. Fawcett, *Chem. Eng. News*, **30**, 2588 (1952).
4. J. F. Voegelein, *J. Chem. Educ.*, **43**, A151 (1966).
5. D. A. Pipitone and D. D. Hedberg, *J. Chem. Educ.*, **59**, A159 (1982).
6. N. V. Steere, *J. Chem. Educ.*, **41**, A859 (1964).
7. R. D. Coffee, *J. Chem. Educ.*, **49**(6), A343 (1972).
8. D. N. Treweek, *Ohio J. Sci.*, **80**(4), 160 (1980).
9. R. Powers and P. Redden, *J. Chem. Educ.*, **59**, A9, (1982).
10. S. H. Pouliot, *A Program for Compatible Chemical Storage of Chemicals*, Thesis, University of North Carolina, 1973.
11. *Chemical Hazard Response Information System (CHRIS) Manual*, 2nd ed., U.S. Coastguard, Washington, DC, 1979.
12. Reference 5, p. A161.
13. *Hazardous Chemicals Data, NFPA 49*, 3rd ed., National Fire Protection Association, Boston, MA (1975).
14. R. C. Weast and M. J. Astle (Eds.), *Handbook of Chemistry and Physics*, (new editions annually), CRC Press, Boca Raton, FL.
15. M. Windholz (Ed.), *The Merck Index*, 9th ed., Merck Company, Rahway, NJ, 1976.
16. L. Bretherick (Ed.), *Hazards in the Chemical Laboratory*, 3rd ed., Royal Society of Chemistry, London, 1981.
17. *Handling Chemicals Safely, 1980*, 2nd (English) ed., Dutch Association of Safety Experts/ Dutch Chemical Industry Association/Dutch Safety Institute, Amsterdam, 1980.
18. *Manual of Hazardous Chemical Reactions NFPA 491M*, 5th ed., National Fire Protection, Boston, MA, 1975.
19. L. Bretherick, *Handbook of Reactive Chemical Hazards*, 2nd ed., Butterworths, Boston, MA, 1979.
20. N. I. Sax, *Dangerous Properties of Industrial Materials*, 5th ed., Van Nostrand–Reinhold, New York, 1979.
21. M. J. Pitt, *Chem. Ind. (Lond.)*, **1982**, 804.
22. Static Hood from Bel Art Products, Pequannock, NJ 07440. Construction details for a

storage cabinet have been published: Di Berardinis, L. *et al., Am. Ind. Hyg. Assoc. J.,* 1983, **44**(8), 583–588

23. K. Everett and D. Hughes, *Guide to Laboratory Design,* Butterworths, London, 1975, pp. 104–111.

24. D. W. Shive and W. H. Norton, *Hazardous Chemical Safety* (course notes), J. T. Baker Chemical Co., Phillipsburg, NJ, 1980, pp. 16.1–16.8.

25. *Prudent Practices for Handling Hazardous Chemicals in Laboratories,* National Research Council, Washington, DC, 1981, pp. 215–229.

26. N. V. Steere, *Handbook of Laboratory Safety,* 2nd ed., CRC Press, Cleveland, OH, 1971, pp. 179–199.

27. M. E. Green and A. Turk, *Safety in Working with Chemicals,* MacMillan, New York, 1978, pp. 16–17, 35–39, 117–119.

28. Reference 26, pp. 574–578.

29. W. Braker and A. L. Mossman, *Matheson Gas Data Book,* 6th ed., Matheson Gas Products, East Rutherford, NJ, 1981.

CHAPTER **3**

LABELING UNSTABLE CHEMICALS

L. JEWEL NICHOLLS

University of Illinois at Chicago, Chicago, Illinois

3.1 INTRODUCTION

Label and storage practices are vital. Labels put onto packages at the factory attest to the contents of that package as it is packaged. Quantities of impurities as small as parts per million (ppm) are reported as in the part of a label shown in Figure 3.1. How long can one be certain of that purity? What effect does standing, heat, air, and moisture have on a particular chemical?

Many companies are improving labels of fine chemicals by placing warnings, dates, and first aid information on materials (see Appendixes), but there are countless bottles of chemicals of unknown age and history in storerooms and laboratories with little information other than a trade name and company name. The wary chemist will consider potential sources of contamination and effects of degradation before using any chemical. Use of opened bottles of substances poses problems of unknown impurities that may have formed or been added. However, to open a new bottle of chemical each time there is a doubt is unthinkable because of the expense of the new supply in addition to the difficulty and expense of disposal of the less certain unused portions of the shelf bottle. Indeed, it becomes a moral problem.

There are certain storage procedures that can be used to minimize contamination. Tight caps and sealants such as wax, tape, or film about caps can decrease seepage of air and vapor in and out of the packages. Storage under an inert atmosphere or in a desiccator can eliminate or minimize contact with air, moisture, or other vapors. Methods of charac-

Formaldehyde solution fw. 30.027	
Analysis of Lot 12345	
Assay (HCHO)	36.9%
Residue after ignition	0.002%
Acidity (as HCOOH)	0.012%
Chloride (Cl)	0.0002%
Sulfate (SO_4)	0.001%
Heavy metals (as Pb)	0.0002%
Iron (Fe)	0.0001%
Preservative (methanol)	11.2%

FIGURE 3.1 Labels on fine chemicals show assays and analyses of impurities found at the time the material is packaged.

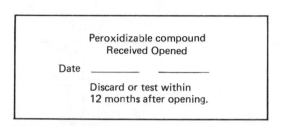

Peroxidizable compound
Received Opened

Date _____ _____

Discard or test within
12 months after opening.

FIGURE 3.2 A label of an unstable class of chemicals that form explosive peroxides. Dating receipt, opening, and shelf life are essential.

terization and purification are well documented. Experimental procedures almost always document methods for purification or characterization. In a synthesis of anhydrous copper(II) nitrate, for example, ethyl acetate must be dried by distillation over P_4O_{10} before use (1). The *Aldrich Catalog* cites an infrared (IR) reference for those chemicals that are unstable (2).

There are labeling practices chemists can use as reminders of the age and potential hazards or impurities. A mark with permanent black ink (e.g., India ink) on the label of any chemical of the date received and date opened is a minimum necessity. Figure 3.2 shows the type of label recommended for peroxidizable materials.

Chemicals that are synthesized or rebottled need to be labeled as completely as possible with durable ink, attached with a firm adhesive, and protected with a layer of lacquer or plastic film. Figure 3.3 shows such an example. In our undergraduate laboratories at the University of Illinois at Chicago, a rectangle of clear adhesive film (available in hardware stores) is cut to about 10% larger than the label. The backing of the film is peeled

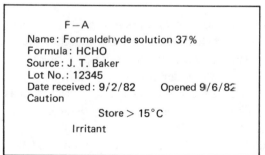

F—A
Name: Formaldehyde solution 37%
Formula: HCHO
Source: J. T. Baker
Lot No.: 12345
Date received: 9/2/82   Opened 9/6/82
Caution

Store > 15°C
Irritant

FIGURE 3.3 A label of a repackaged chemical showing the necessary information. The abbreviation "F-A" represents a storage shelf code for convenient retrieval and filing.

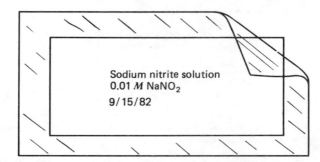

FIGURE 3.4   Protection of a paper label with transparent plastic film.

off, a label is attached to the adhesive, and the assembly is attached to the bottle as shown in Figure 3.4. The procedure is efficient, making the label durable. The contents of the bottle and pertinent information such as formula, source, lot number, date received, date rebottled, storage information, hazards, and expiration date may all be necessary for future use. Loose labels are covered and secured by this means as well.

3.2   CHEMICALS AND TIME-CAUSED DEGRADATION

Thermodynamically unstable compounds will eventually break down into simpler molecules or elements. The Cannizzaro reaction is the base-catalyzed decomposition of aldehydes to yield corresponding alcohols and acids as in Eq. 3.1.

$$2CH_2O + OH^- \rightarrow CH_3OH + HCOO^- \tag{3.1}$$

The water present as solvent or impurity provides the base over a long period of time (3), so that instead of the reaction the text describes as occurring in an hour or two with strong base, it reacts over a period of 1 year or more in the bottle under very mild conditions.

Many aldehydes and ketones condense to form aldols as in the reaction of an aldehyde to give a high-molecular-weight solid depicted by Eq. 3.2.

$$2(CH_3)_2CHCHO \rightarrow (CH_3)_2CHCHOHC(CH_3)_2CHO \tag{3.2}$$

The equilibrium established is disturbed in some cases by further reaction to dimeric aldols or alpha, beta-unsaturated aldehydes. These reactions can be catalyzed by acids or bases (4). Bottles originally containing colorless liquid aldehydes after long storage have been observed to hold dark solids. Thus the physical properties of the substance or standing should be a clue to purity.

3.3 EXPLOSIVES: SHOCK-SENSITIVE COMPOUNDS

It is well known that noncompatible compounds such as oxidizing and reducing agents react violently when mixed together. An example is a mixture of a reducing agent such as hydrazine and an oxidizing agent such as nitric acid as expressed in Eq. 3.3.

$$N_2H_4 + HNO_3 \rightarrow 1.5N_2 + 2.5H_2O + 0.5O_2 \qquad (3.3)$$

The same potential exists in molecules that have a reducing part and an oxidizing part. A well-known potentially explosive chemical is picric acid or trinitrophenol. The aryl part of the molecule, like phenol, is toxic and combustible and is oxidized at elevated temperatures. In the decomposition of picric acid, the nitro groups have an abundance of oxygen which combines with both hydrogen and carbon of the aryl ring to form stable gases. The nitrogen atoms combine to form highly stable nitrogen gas. Decomposition of picric acid can be shown (5) to be most energetically favorable according to the reaction shown in Eq. 3.4.

$$C_6H_3N_3O_7 \rightarrow 1.5H_2O + 5.5CO + 0.5C + 1.5N_2 \qquad (3.4)$$

Thus picric acid may decompose to yield more energy and gaseous byproducts than TNT. The temperature of the gases in an adiabatic decomposition is calculated to be 2464°K (T_d). The temperature is even higher when the flame reacts with oxygen. This flame temperature is calculated to be 3051°K (T_o). Table 3.1 compares thermodynamic properties of several common chemicals.

Organic peroxides are a class of compounds even more shock sensitive than explosives such as TNT or picric acid. Even small amounts of these are not stored under long-term normal conditions because of the tendency to decompose both slowly and uncontrollably. Small amounts must be refrigerated for the short time before use and then destroyed adequately

TABLE 3.1 Enthalpy and Final Temperature of Explosion and Combustion of Selected Compounds

Compound	ΔH_d (kcal/100 g)	ΔH_o (kcal/100 g)	T_d (°K)	T_o (°K)	V_d (liters/100 g)
Ammonium nitrate NH_4NO_3	-35	-35	1246	1246	358
Trinitrotoluene $C_7H_5N_3O_6$	-63	-165	2025	3109	1246
Picric acid $C_6H_3N_3O_7$	-72	-130	2464	3051	1819
Acetic acid CH_3COOH	-1	-200	302	2771	50

Note: ΔH_d—enthalpy of decomposition in a reaction that yields simple compounds or elements to give the most energy; ΔH_o—enthalpy of combustion per mole; T_d—adiabatic flame temperature of decomposition in absence of air or oxygen; T_o—adiabatic flame temperature in the presence of oxygen; V_d—volume of gaseous products of the reaction at T_d per 100 grams of reactant.

before disposal. Amounts over 25 g are normally stored only in specially constructed magazines.

The *Aldrich Catalog* (2) indicates that dibenzoyl peroxide, $(C_6H_5CO)_2O_2$, cannot be shipped by Parcel Post or United Parcel Service (UPS) in the 50- or 500-g sizes. The 70% solution of dibenzoyl peroxide in water is offered, of which the 100-g size may be shipped. Comparison of the thermodynamics of peroxides and other organic compounds indicates that the energy of activation of explosion is substantially less for peroxides than for other organic compounds. Table 3.2 lists a few values for typical compounds (6).

Not all explosives are organic molecules. Ammonium nitrate decomposes explosively as Eq. 3.5.

$$NH_4NO_3 \rightarrow N_2 + 2H_2O + 0.5O_2 \qquad (3.5)$$

The energy given off is such that the final temperature in an adiabatic decomposition would be 1246°K. The volume of gases produced at this temperature is 358 liters per 100 g. In this case, an amount of unreacted oxygen is present. If a reducing agent such as powdered metal were exposed to the oxygen, even more energy would be released.

Coordination compounds are often prepared by precipitation with a

TABLE 3.2 Thermodynamic Properties of Selected Organic Peroxides

	ΔH_f (kcal/mol)	T_d (°K)	E_a (kcal/mol)	ΔH_c (kcal/mol)
Acetyl peroxide $C_2H_6O_4$	−116.1	983	29.5	254
Peracetic acid $C_2H_4O_3$	−97.7	976	32.0	206
Acetic acid $C_2H_4O_2$	−104.9	634	67.5	209
Ethyl ether $C_4H_{10}O$	−61.9	761	78.0	658

Note: Organic peroxides have thermodynamic properties similar to other materials except for the lower energy of activation of explosion, E_a, which is indicative of the fragile peroxide linkage that breaks, often explosively, at temperatures only slightly above room temperature; ΔH_f is the molar enthalpy of formation, T_d is explained in Table 3.1, and ΔH_c is the molar heat of combustion. Adapted from Ref. 6.

large oxyanion such as the perchlorate ion. These compounds have the potential for explosion with shock or spark. Hexaaminenickel(II) perchlorate decomposes explosively to give 13 moles of gas per mole of compound that must yield more than 800 liters per 100 g at the temperature of decomposition:

$$[Ni(NH_3)_6](ClO_4)_2(s) \rightarrow 3N_2(g) + 2HCl(g) + 8H_2O(g)$$
$$+ Ni(s) \tag{3.6}$$

Equation 3.6 shows the complex decomposing to the most stable gaseous compounds although the nickel produced in the reaction is assumed to be elemental.

Acetic acid has a higher flame temperature and lower decomposition temperature than most explosives, in keeping with experience that acetic acid is combustible but nonexplosive.

$$CH_3COOH \rightarrow 2C + 2H_2O \tag{3.7}$$

$$CH_3COOH + 2O_2 \rightarrow 2CO_2 + 2H_2O \tag{3.8}$$

Comparison of these reactions indicates the reason. The combustion reaction (Eq. 3.8) yields twice the amount of gaseous product as the decomposition reaction (Eq. 3.7). The greater energy released in the combustion heats up the gases to a much higher temperature, T_0, which expands the gases even more (see Table 3.1).

Chemicals that have potential for explosive reactions should be stocked in small amounts and handled carefully. Large amounts should be treated and stored (7) as explosives. Labels should denote this explosive characteristic. Picric acid has been used so widely in the past that it has been found in the stores of many unsuspecting researchers and educators. People have reacted in panic, calling in bomb squads for disposal. The Department of Transportation requires not less than 10% moisture in picric acid for safe shipment. Under arid conditions of storage this moisture content decreases, increasing the danger. Immersion of the container upside down in water for several days so that the crystals around the lid become moist can reduce the danger of spark-initiated detonation on opening. Addition of water to the chemical at regular intervals will reduce sensitivity toward shock or friction.

Several structural and thermodynamic properties may contribute to the explosive decomposition of many substances. Of these, the heat and volume of decomposition and the structural features give a clue. Chemical names that contain *per-*, *peroxy-*, *azo-*, and *acetylide* can alert one to the fragile bonds of peroxides, azides, and acetylides. A large amount of oxygen in an organic molecule may mean a large volume of gas and great energy released on decomposition. This may occur for nitro groups, perhalates, halates, or halites.

3.4 AIR- AND MOISTURE-SENSITIVE CHEMICALS

Many compounds react with air, moisture, or impurities to change composition in normal storage. Such processes may be enhanced by application of heat or light and by presence of trace impurities that may catalyze reactions or decomposition.

Ethers and some other compounds react with air to form peroxides, so it is imperative to mark the date received, the date opened, and the expected shelf life (8) as shown in Figure 3.2. These peroxides are less volatile than the solvent itself and tend to concentrate in a flask during a distillation where the heat applied to a crystal or concentrated solution of the peroxide can trigger an explosive decomposition. Ethyl ether is not the only compound to have this hazard. Tetrahydrofuran, dioxane, diisopropyl ether, and methoxy and ethoxy compounds of low molecular weight are all peroxidizable to varying degrees. Many such compounds are packaged with oxidation inhibitors such as hydroquinone. Testing susceptible compounds and replenishment of inhibitors as they are used up is a way to extend the shelf life of the compound. An excellent review of structure types, limits of storage, conditions of storage, and procedures for testing and removal of peroxide from many peroxidizable compounds has been published in the *Journal of Chemical Education* (8) (see Table 3.3).

It is dangerously naive to assume that the label, "Tetrahydrofuran spectroanalyzed, no preservatives" means that the substance is so pure that it can be used for many purposes and that it will stay as pure as it states. Experiments have determined that a percentage of 0.008% or more of peroxide (tested as (H_2O_2) in any compound can be dangerous, but according to the catalog of a respected supplier, none of the THF offered with or without preservatives is guaranteed to have less than 0.015% peroxide (as THF peroxide). Thus newly opened containers (as well as older ones) of peroxidizable compounds must be tested before using as a solvent for distillation or refluxing. There are test papers available for such tests, but it is easy to keep a brown dropper bottle of 10% potassium iodide for routine testing.

Certain compounds, most notably diisopropyl ether, decompose rapidly on storage so that crystals of peroxide accumulate in the solution or line the cap of the bottle. This condition is extremely dangerous, and such containers should not be opened. Handling, if at all necessary, must be done very gently. A bomb squad from the local fire department or army demolition team may be needed to dispose of peroxidizable chemicals in this condition. Remote detonation in an open field has been employed by bomb squads for ultimate destruction.

Peroxide initiation of polymerization is a problem with storage of monomers such as butadiene as indicated in Eq. 3.9.

$$n\mathrm{CH_2CHCHCH_2} \rightarrow \mathrm{(CH_2CHCHCH_2)_n} \qquad (3.9)$$

TABLE 3.3   Examples of Peroxidizable Compounds[a]

Red Label—Peroxide Hazard on Storage—Discard After 3 Months

Isopropyl ether
Divinyl acetylene
Vinylidene chloride
Potassium metal
Sodium amide

Yellow Label—Peroxide Hazard on Concentration—Discard After 1 Year

Diethyl ether	Dicyclopentadiene
Tetrahydrofuran	Diacetylene
Dioxane	Methyl acetylene
Decahydronaphthalene (Decalin)	Tetrahydronaphthalene (Tetralin)
	Cyclohexene
Ethylene glycol dimethyl ether	
Vinyl ethers	
Acetal	

Yellow Label—Hazardous to Peroxide Initiation of Polymerization[b]— Discard After 1 Year

Methyl methacrylate	Chlorotrifluoroethylene
Styrene	Vinyl acetylene
Acrylic acid	Vinyl acetate
Acrylonitrile	Vinyl chloride
Butadiene	Vinyl pyridine
Tetrafluoroethylene	Chloroprene

[a]From H. L. Jackson, W. B. McCormack, C. S. Rondestvedt, K. C. Smeltz, and I. E. Viele, "Safety in the Chemical Laboratory. LXI. Control of Peroxidizable Compounds," *J. Chem Ed.*, **47**(3), A176 (March 1970). Reprinted with permission.

[b]Under conditions of storage in the liquid state the peroxide-forming potential increases and certain of these monomers (especially butadiene, chloroprene, and tetrafluoroethylene) should then be considered as Red-Label compounds.

3.5   CATALYTIC EFFECT OF LIGHT AND IMPURITIES

In addition to acid- and base-catalyzed reactions, many photochemical reactions of carbonyl compounds have been observed (9) as illustrated in Eq. 3.10. These fragments combine to give impurities. Therefore, storage in dark bottles away from sunlight is essential.

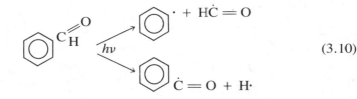

$$(3.10)$$

Inorganic chemicals may also decompose. Some oxidation states of elements disproportionate to more stable mixtures. One can use oxidation potentials to predict stabilities, as in unusual oxidation states of metal ions such as manganese(III), which is shown in Eq. 3.11.

$$2Mn^{3+} \rightarrow Mn^{2+} + Mn^{4+} \qquad (3.11)$$

Thermodynamically unstable compounds decompose to more stable compounds or elements. Using standard free energies of formation one can predict or confirm reactions of decomposition such as that of hydrogen iodide shown in Eq. 3.12.

$$2HI \rightarrow I_2 + H_2 \qquad (3.12)$$

The free energy of formation of aqueous HI is -12.38 kcal/mol. However, in the gas phase the free energy is positive. Therefore, HI should be relatively stable in aqueous solution, but not in the gas phase. The vapor over a concentrated solution thus decomposes, which is accelerated by the presence of light. This phenomenon was observed in a distorted, discolored plastic bottle that was labeled "Conc. [concentrated] HI" depicted in Figure 3.5. Apparently when the above decomposition had taken place, the hydrogen molecules produced were small enough to diffuse through the plastic, yet the resulting iodine accumulated in large amounts inside the bottle. The concentrated reagent had been originally packaged in an amber-colored glass bottle. The amount used in the laboratory should have been repackaged in brown glass to prevent the photodecomposition, diffusion, and probable reaction with plastic by elemental iodine.

   In the case of hydrogen peroxide, however, the oxygen produced through decomposition causes a pressure increase in plastic bottles that must be vented. A plastic bottle of 30% H_2O_2 stored on a shelf at ambient temperature was seen rocking on its convex bottom as in Figure 3.6 because the

FIGURE 3.5 Effect of light on decomposition of concentrated hydrogen iodide solution stored in a clear plastic bottle, resulting in discoloration and distortion of the bottle.

cap could not release the built-up pressure. This is an example of a compound that will decompose rapidly with heat and light and in the presence of impurities. Packaging of chemicals in bottles with metal screw caps would cause not only rusted caps but dangerous build-up of pressure.

3.6 STORAGE OF CORROSIVES

Corrosive vapors have adverse effects on shelves and storage areas. Following observation, the author concluded that one shelf had stored bromo

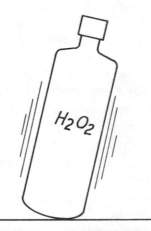

FIGURE 3.6 A deformed plastic container of 30% hydrogen peroxide, caused by unvented pressure. The presence of a metal cap can catalyze such a decomposition.

compounds because of the rust on the back and sides of the painted metal shelves as in Figure 3.7. Opened or old bottles had spilled hydrogen bromide and leaked vapors of the contents. This is true of acid chlorides as well. Sealing the containers with wax or parafilm and disposing of bottles with damaged caps corrected not only the corrosion problem, but also the stench associated with that area of the storeroom.

Rings on shelves caused by leakage are eyesores and also indicate which containers are faulty. Placement of the container in a plastic bag is no remedy, since the label is then destroyed by the leak. Repackaging may be necessary, but loss of the original label renders the product less useful and less desirable. Just wiping up the ring and ignoring the problem solves nothing.

Metal cans that serve as packing containers are easily rusted and should not be used as shelf packs. The author noticed oil on a shelf one day near a rusty container of yellow phosphorus and discussed the problem with the storeroom personnel, who then prepared an immersion bath of water. Before it could be immersed, the phosphorus began to smoke, and the oil in which the phosphorus was packed ignited in flames. Using tongs, we put the container in the hood where the heat of the reaction melted the can and broke bottles. We were fortunate in that no one was hurt and no damage done. An amusing aspect of the cleanup was that every time we scraped a bit of phosphorus off the wall of the fume hood, it was like

FIGURE 3.7   Telltale signs of corrosive chemicals: rusted shelves and spotted walls, indicating the escape of corrosive vapors.

striking a match. This is a case in which repackaging was necessary. Since the inner can had not been visible, we could not tell that it had failed.

Perchloric acid can be particularly corrosive in storage. Experts suggest storage of concentrated reagent only in perchloric acid-designed hoods in secondary glass trays that will hold all contents. Grave danger occurs when the perchloric acid evaporates and condenses on the walls and shelves in the area. A mixture of perchlorate and any organic compound is explosive and can be touched off by friction such as adjusting the panels in a hood.

3.7 STORAGE PRACTICES

In view of these uncertainties, the prudent chemist must take several precautions.

1. Place one date on the label when any chemical is received and another when opened. In addition, an expiration date should be added that gives the user definite warning if the substance is old and indicates that the material should be tested after that period of time.

2. If any doubt exists regarding the content or purity of an organic compound, an IR spectrum can confirm its purity or give a clue to the kinds of impurity.

3. Warn users of susceptible solvents to test ror peroxide content before heating.

4. Any chemical that can form peroxides should be handled very carefully and not opened at all if:
   a. It is of uncertain age.
   b. It has formed solid particles.
   c. Its physical characteristics differ from those of the pure substance.

5. Treat potentially explosive compounds gently and place a warning on the label. Store large quantities in special facilities.

6. Seal all containers well.

7. Secure labels. Once the label is lost your work really starts.

8. Dispose of aged and decomposed compounds on a regular basis.

REFERENCES

1. R. J. Angelici, *Synthesis and Technique in Inorganic Chemistry*, Saunders, Philadelphia, 1977, p. 102.

2. *1982–1983 Aldrich Catalog*, Aldrich Chemical Company, Milwaukee, 1982.

3. C. K. Ingold, *Structure and Mechanism in Organic Chemistry*, Cornell University Press, Ithaca, NY, 1969, p. 1029.

4. Ibid., pp. 999–1003.

5. D. R. Stull, *J. Chem. Ed.*, **48**, A173–A182 (1971).

6. S. W. Benson and H. E. O'Niel, "Kinetic Data on Gas Phase Unimolecular Reactions," NSRDS-NBS 21, Superintendent of Documents, U.S. Government Printing Office, Washington, DC, 1970; through D. R. Stull, *J. Chem. Ed.*, **51**, A21 (1974).

7. NFPA 495-1973, Code of Management, Transportation, and Storage of Explosives or Section 1910.109 in the *Federal Register* (125), 1974.

8. H. L. Jackson, W. B. McCormack, C. S. Rondestvedt, K. C. Smetz, and I. E. Vielle, *J. Chem. Ed.*, **47**, A175–A188 (1970).

9. J. N. Pitts and J. K. S. Wan, in S. Patai (Ed.), *Chemistry of the Carbonyl Group*, Interscience, New York, 1969, pp. 897–898.

CHAPTER 4

COUNTERACTING SPILLS
IN THE CHEMICAL STOREROOM

DAVID A. PIPITONE

Technical Director, Lab Safety Supply Company, Janesville, Wisconsin

4.1 INTRODUCTION

Spill control has become an increasingly popular topic. At least three major books and several articles on the topic of spill response have been published within the past 4 years (1–4). Newspaper and television headlines tell of a wrecked truck here, or a derailed tanker there, with the resulting evacuations and injuries. Chemical storerooms do not have the problems associated with massive releases of hazardous chemicals to the same degree as do transportation accidents. Discharge to the environment may happen on a limited scale, as when a spill from a leaking drum may enter the sewer. Yet with the advent of federal hazardous waste regulations, chemical storerooms, in addition to the daily stock of unused hazardous chemicals, have the burden of storing bulk quantities of hazardous waste. A chemical spill can pose an immediate threat to the health and life of stockroom personnel. A spilled hazardous chemical represents the danger of exposure to toxic and corrosive material, potential for fire, as well as an excruciating slip on a wet surface. Fortunately, proper planning can result in an effective program for spill prevention.

4.2 SPILL CONTROL AND COUNTERMEASURES PLAN (SCCP)

The possession and use of a spill control and countermeasures plan (SCCP) for a chemical storeroom is essential to inform both stockroom personnel and response team members of a twofold program. The plan establishes practices to avert accidental spillage of chemical through management controls and serves as a guide for personnel responding to a spill scene. Spill *prevention* consists of determining the possible causes of spills and taking corrective action. Spill *response* depends on knowledge of the spilled material, accurate choice of response equipment and gear, and trained personnel to use established procedures to clean the area and dispose of the spent material in a proper fashion.

4.2.1 Preventive Management Measures

Through appropriate measures, relating to stockroom operation, accidental spills can be avoided through preplanning.

Spills can occur by (a) rupture of chemical containers, (b) inadequate shelving space, (c) inadequate shelving integrity, (d) lack of guards on shelves, and (e) inappropriate handling. With the use of management controls, the causes can be eliminated or lessened so that the risk of spills is reduced.

Chemicals can be released by the rupture of their container. Containers that have aged, in the evidence of rusted metal cans, lack the structural integrity once theirs. Overpressurized containers, as evidenced by the deformation in container shape, have the potential for rupture and chemical spills. Remedies for leaking and overpressurized 55-gallon drums have been discussed in the literature (5). A regular inspection of the chemical stock will reveal those containers that are leaking or have the potential to leak. Repacking the chemical into another container must be accomplished promptly. If the chemical purity is suspect, disposal of both chemical and container is in order.

The shelving system can be the cause of a major chemical spill. Inadequate shelving space that results in overcrowded shelves may well surpass the rated weight limit of the shelving. A prudent measure for remedying this problem is to obtain the weight limit of a shelf from the manufacturer and determine the weight of chemicals (from the containers) that can be placed on the shelf. Shelving units that are not braced or secured to the wall or floor may collapse without notice, or, even more dangerously, fall on personnel reaching for a chemical. A few fasteners to secure the shelving units can save much trouble.

Chemicals tend to "creep" toward and over the edge of a shelf (6). Provision of a raised lip on a shelf 1/4 inch above shelf level is an economical and simple procedure that can prevent an unexpected spill. Installation of such a lip is demonstrated in Figure 4.1.

Inappropriate handling or transporting of chemicals can result in a dropped, broken, and leaking container. Solutions for safe transport of these containers include the use of carts and bottle carriers or pails. Some chemical suppliers now package chemicals in glass bottles that are encapsulated in a plastic film. If the bottle is dropped and breaks, the plastic film acts as an "envelope" to hold the liquid inside. These bottles are also commercially available, so that if repackaging of a chemical is needed, as indicated above, the same benefits may apply.

Smooth operating procedures, material handling, and scrutiny of the storeroom contents are the best management control for spill prevention. Establishing a routine inspection on a regular basis can result in the

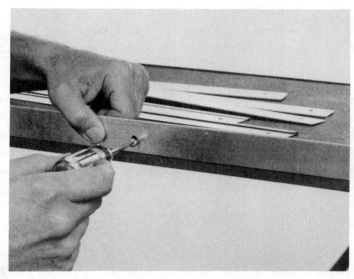

FIGURE 4.1 Installation of raised edges on shelves. A nominal height of ¼ inch is required. (Photograph courtesy of Lab Safety Supply Company, Janesville, Wisconsin.)

TABLE 4.1   Inspection Checklist for Preventing Causes of Spills

Yes	No	
		Metal containers are free of rust
		Containers are clean and free of any chemical leakage
		Containers shapes are free of the signs of pressure build-up
		Container closures are secure and free of deformation
		Shelving units are fastened to the wall and/or floor
		Shelves are free of overcrowding
		Weight limits for shelves are posted
		Weight limits for shelves are not exceeded by the weight of chemicals
		Raised edges are an integral part of each shelf
		Shelving units are braced
		Chemicals are transported using carts, bottle carriers, or pails

reduction of accidental spillage. The checklist in Table 4.1 may be adopted for this purpose.

4.2.2   Planning Emergency Response to Chemical Spills

Spills in the storeroom all represent the potential for uncontrolled release and discharge of a chemical. Some chemicals may pose minimal hazards, if at all, when spilled. The danger results from the wide variety of hazardous chemicals (Table 4.2) that require different response tactics rising from the nature of the hazards. Lee (7) indicates five areas of concern when facing a spill that confront the personnel: (a) lack of information about the hazard; (b) lack of information about resources; (c) lack of tactical information; (d) lack of response capability; and (e) lack of training. Since a spill may occur suddenly, having an informed plan is the first measure for emergency response. A useful plan, then, will (a) inform personnel of the hazards, and current resources for cleanup, (b) provide tactics and capabilities for response, and (c) schedule and maintain training.

The scope of an emergency response plan may vary widely and will depend on the inherent resources possessed by the size and sophistication of those attending the storeroom, in addition to the nature of the chemicals

TABLE 4.2   Examples of Hazardous Chemicals for Which a Spill Response Plan is Necessary

Liquids	
Acids	Nonreactive salt solutions
mineral	Oxidizers
oxidizing	Mercury
organic	Organic peroxides
Caustic	Water-reactive
Flammables	Pyrophoric
Toxic	
Reducing agents	
Solids	
Oxidizers	Caustic
Water-reactive	Toxic
Pyrophoric	

spilled. A small school that stores chemicals and has limited funds and personnel would be well advised to rely on a trained, well-equipped response team from a local fire department for spills of toxic chemicals or large releases of flammable liquids. A large institution may have an organized response team promptly ready on notification. Whatever the case, a safe cleanup should be the immediate goal.

A plan must be based on an accurate source of information for response to each chemical or group of chemicals. This must incude the hazards of the chemical, the type of equipment available for cleanup, and its location, procedures for cleanup, personnel protection, first aid information, and waste disposal procedures. Hazards of the chemical include toxicity, flammability, reactivity with air or water, the corrosive nature of the chemical, and the physical state. A liquid spill can be more dangerous than a solid spill because liquids can vaporize and flow under shelves or doors or across aisles.

Hazard information is available from a number of sources (8–14). The *CHRIS Hazardous Chemical Data* (8) developed for the U.S. Coast Guard for response to water spills of hazardous materials and the *NIOSH–OSHA Pocket Book to Chemical Hazards* (11) give a summary of health, flammability, and reactivity hazards for a large number of chemicals. *Toxic and Hazardous Chemicals in Industry* (10), *Handling Chemicals Safely* (9), and *Dangerous Properties of Industrial Materials* (12) also provide hazard and handling data. A breakdown of the applicable information will follow in the next section. To have a complete system of information on all the hazardous chemicals in stock, one must take an inventory of those chemicals, note which chemicals require special handling, and develop an appropriate data file for spill response purpose.

The type and amount of cleanup equipment available will depend on the chemical spilled. A solid spill will require a means of response different from that provided by a liquid spill. Corrosive liquid acids may be treated differently from when cleaning a spill of a flammable liquid. The size of the spill will affect the amount and type of absorbents, neutralizers, or mechanical equipment needed for efficient response. The toxicity of the spilled material can cause the use of different apparatus. Section 4.3 discusses the use of such equipment at length.

Procedures for cleaning a spill will depend on the location and nature of the spill. Procedures for neutralization differ from those of absorption. Procedures for cleaning up a solid spill differ from those for a liquid spill.

It is well to have noted in the plan differences in procedures that reflect the effects of concentration of the same material, as for dilute and concentrated acids. A written procedure that can be retrieved as an instruction sheet in the time of an emergency will do much to assist the response team.

Personal protective equipment must be worn before entering the scene of a spill. What kind must be worn, and for which chemical? A plan that researches the characteristics of appropriate protective clothing, eyewear, gloves, and respiratory protection required for handling a spill emergency of a specific chemical and adds this information to the data file answers this question. The amount of protective gear required will again depend on spill size, method of treatment (e.g., neutralization of an acid will cause splattering of acid), and concentration of the hazardous properties (toxicity, corrosivity).

First aid information is important in the event of an injury or chemical exposure. The possession of informed manuals (10,15,16) with detailed first aid procedures for each chemical is a must. It is essential that this information be easily retrieved by response personnel. Additional information for medical treatment should be readily available.

Information for packaging the spill debris and waste disposal must be known so that any resulting hazardous material is properly containerized and promptly discarded. Awareness of Federal Regulations (17) for the proper disposal of hazardous waste [including spill residue or debris of those chemicals listed in 40 CFR of Section 261.33 (e),(f)] is essential. Since spill debris will be stored until disposal, packaging in leakproof containers is essential.

The discussion presented above points to a systematic information system that can give the needed facts regarding the occurrence of an emergency. Reliance on human memory can lead one to omit important details. The sophistication of such an information system depends on the need and inventiveness of the individual storeroom. A reference library will suffice for some, but a system that is organized by type of chemical will be the most helpful. Development of a tailor-made database by using notebook or computer files may seem cumbersome or difficult but is very rewarding in the case of an emergency. Table 4.3 shows the functional elements of such a conceptual database. This format may resemble a material safety data sheet; however, the reader should note specific references to room, location, type of absorbents, and so on. The content should be designed

TABLE 4.3 Conceptual Spill Response Plan for Acetone

Chemical:	Acetone	Location:	Room B1297
Class:	Flammable liquid		Room B1298
			Room B2397

Number/Size of Containers

20/4-Liter bottles
3/5-Gallon drums
1/55-Gallon drum

Hazards

Health: TWA 750 ppm
PEL 1,000 ppm
IDLH 20,000 ppm
Avoid skin contact
Flammability: Flash point 0°F (C.C)
LFL, 2.6%; UFL, 12.8%
Flashback may occur
Reactivity: Slight
Incompatible with strong oxidizers

Spill Response Equipment/Location

1. Absorbent pillows/center stores, closet 2—one pillow per gallon spilled
2. Diatomaceous earth—nonsparking shovel/Room B1263—spread until liquid is absorbed
3. Empty 17H drums/Room B1263
4. Barricade tape/Room B1263

Protective Clothing/Location

1. Disposable jumpsuits/center stores, Closet 1, Shelf 1
2. Butyl rubber gloves/center stores, Closet 1, Shelf 2
3. Self-contained breathing apparatus/center stores, row 1

for quick reference and extraction of the needed information. The database must be practical and reliable to work properly. The rest of this chapter discusses the selection and use of those functional elements that can "fill the data blanks" in the spill response file.

4.2.3 Chemical Hazard Information

The need for specific information on chemical hazards that relate to a stockroom spill is the first consideration when cleaning the spill of a

TABLE 4.3   (*Continued*)

Fire Response

Use carbon dioxide or Halon 1211 fire extinguisher

Spill Response Procedure

1. Barricade spill scene from traffic and evacuate nonessential personnel
2. Notify safety officer
3. Contact response team: Bill (Ext. 3243), Joel (Ext. 3246)
4. Don protective clothing and review hazards
5. Remove all ignition sources
6. Dike spill and treat liquid with absorbent
7. Shovel debris into drum
8. Scrub and ventilate spill area
9. Seal drum and label
10. Report spill cleanup details
11. Replenish spill response equipment

First Aid

Inhalation: Remove to fresh air; if breathing has stopped, give artificial respiration; if breathing is difficult, give oxygen
Eye exposure: Flush eyes for a minimum of 15 minutes, holding eyelids open

Waste Disposal

Spill debris is EPA hazardous waste (D0001—ignitable)
DOT Label: "Flammable solid"
Place debris in DOT Spec 17H metal drum; send drum to approved landfill or incinerator

---

particular chemical. The individuals who effect this cleanup have the potential for exposure to hazardous chemicals, as do the immediate surroundings. A summary of the hazards that must be identified is given below, with references.

Health Hazards.   Inhalation—TLV, IDHL, LC_{50} (9–11,18); saturated vapor concentration, skin exposure—TWA, LC_{50} (10,18), corrosivity to skin (9,10); absorption through skin (8,10), toxicity by ingestion (8,10).

Fire Hazards.   Flash point (8–11), special hazards of combustion products (8,10), upper/lower flammability limits (8–10), type of extinguisher needed (8–10), potential for flashback (8), ignition temperature (8–11).

Reactivity Hazards.   Water-reactive (4,8–14), pyrophoric (8–10,14).

Physicochemical Properties.   Physical state—20°C (8–14), vapor (gas) density (8–14), specific gravity (8–14), evaporation rate (19).

The information should be obtained and recorded for each hazardous material in the stockroom. A hazardous material can generally fall into one or more of the following categories (20):

Acid	Water-reactive chemicals
Caustic	Pyrophoric chemicals
Flammable liquids	Poisons (solid or liquid)
Oxidizers	Reactive chemicals

A list of inventory may be compared against these general classes of chemicals to obtain the names, locations, and amounts of hazardous chemicals in stock.

The use of Material Safety Data Sheet (MSDS, OSHA 20 Form) that has been completed for a specific chemical will be an easier route than checking the references listed. The completeness of the MSDS must be thorough. Many data sheets currently supplied by manufacturers have a limited amount of hazard information and are of little help in preparing a spill response data file. The use of only an MSDS for spill response data, no matter how complete, limits the spill response plan to complacency. Unless a plan is designed for each institution, the workability is suspect.

Attention must be called to the saturated vapor concentration. Since spills in a storeroom occur *inside* a building, the concentration of vapors in an area around and over the spill of a volatile material represents a danger. A "pocket" of toxic or corrosive gases or vapors can cause unsuspecting personnel to be promptly overcome with adverse effects. Pitt (21), in discussing a Vapor Hazard Index (VHI), calculates the relative hazard involved with the accumulation of vapors by ratio of saturated vapor concentration to the threshold limit value. Table 4.4 includes a listing of saturated vapor concentrations for some common inorganic and organic liquids. One may use these numbers as a guide to estimate what concentrations may occur over time when a liquid has been spilled in an unventilated area. The vapor concentration near a recently spilled liquid will depend on the rate of evaporation and the time elapsed since the spill occurred. These values are important when accessing what type of respiratory protection is required.

TABLE 4.4 Comparison of Saturated Vapor Concentration to Health Values[a]

	Saturated Atmospheres (ppm)	TLV[18]	STEL[18] (ppm)	IDHL[11] (ppm)
		Inorganic		
Ammonium hydroxide, 35%	1,000,000	25 ppm	35	500
Ammunium hydroxide, 25%	500,000	20 ppm	35	500
Bromine	230,000	0.1 ppm	0.3	10
Formalin, 40%	1,600	2 ppm (C)	2 (C)	100
HCl, 36%	140,000	5 ppm (C)	5 (C)	100
HCl, 20%	250	5 ppm (C)	5 (C)	100
Hydrofluoric acid, 70%	151,000	3 ppm	6	20
Hydrofluoric acid, 50%	18,000	3 ppm	6	20
Mercury	260,000	0.05 mg/M^3	—	
Nitric acid, 70%	4,000	2 ppm	4	100
		Organic		
Acetone	350,000	750 ppm	1,000	20,000
Benzene	98,000	1 ppm	25	2,000
2-Butanol	17,000	100 ppm	150	10,000
Carbon tetrachloride (skin)	120,000	5 ppm	20	300
Chloroform	210,000	10 ppm	50	1,000
Ethyl acetate	100,000	400 ppm	—	10,000
Heptane	52,000	400 ppm	500	4,250
Methanol	127,000	200 ppm		25,000
Pyridine	27,000	5 ppm	10	3,600
Toluene (skin)	29,000	100 ppm	150	2,000
Xylene	12,000	100 ppm	150	10,000

[a]Possible "pocket" concentrations of vapors that may occur when saturated atmospheres are reached as a result of evaporation of spilled liquid.

4.3 METHODS OF SPILL CLEANUP

The method to be used for spill cleanup depends on the material spilled and the size, toxicity, and reactivity of the material The various types of spills, and alternative methods, are discussed below.

4.3.1   Solid Spills

Spills of solid chemicals are easier to clean than those involving liquids for three reasons: (a) solid chemicals are normally packaged in smaller containers; (b) spills of solid chemicals do not flow uncontrollably as do liquid spills; and (c) solid spills can be physically removed from the floor or shelf without much special equipment. Hazardous solid spills can, however, demand more careful attention because of the highly concentrated form, and sometimes toxic, flammable, or reactive nature. Solid spills can normally be swept by a broom into a dust pan and placed into a suitable waste container. Spilled oxidizing solids such as nitrates, permanganates, perchlorates, and so on must not be dumped with combustible materials such as paper. Extremely toxic dusts such as beryllium, cadmium, arsenic compounds, barium, and mercury compounds may be collected by using a HEPA-filtered vacuum cleaner (22). Such a vacuum cleaner has an absolute filter that removes 99.97% of particles that have a mean diameter as small as 0.3 microns.

Spills of white phosphorous evoke danger because the chemical burns when exposed to air. Keeping spilled phosphorus wet and then covering it with wet sand is a preliminary action before recovering the material (23). The spilled residue must be kept under water to prevent ignition. Other pyrophoric materials must be investigated as to their compatibility with water before using this method.

Spills of water-reactive chemicals such as sodium or potassium must be treated completely differently. These chemicals react with water to form flammable gases that can be ignited by the heat produced by the reaction. Covering potassium with dry sodium carbonate and dispersing and incinerating the mixture in a large steel pan located in an isolated place is recommended by one source (24). Sodium and potassium are normally stored under mineral oil to prevent contact with moist air. Recovered material should be placed into a crock containing enough mineral oil to cover the metal.

4.3.2   Hazardous Liquid Spills

Spills of hazardous liquids cause concern for the following reasons: (a) liquids can flow to other areas bringing the hazardous properties along; (b) liquids can emit gases or vapors which can be toxic, flammable, and/or corrosive; and (c) liquids present a greater slipping hazard than do solids. Liquids cannot be mechanically removed from a surface easily without wicking or conversion into a solid state. This may be done by adding an absorbent directly to a liquid spill or a treated liquid spill. The two main "camps" of spill control for liquids are chemical inactivation (neutralization) and absorption (of untreated liquids). Before choosing either method, the reader must understand which method works better for both the chemical and the spill situation.

4.3.2.1 Chemical Inactivation. This method depends on the interaction of two chemicals to result in a harmless, third chemical that may be easily recovered and discarded. Commonly referred to as neutralization, this method has been traditionally used for response to spills of inorganic acids and bases, although Hedberg indicates the application to other chemicals (25). Addition of a weak base (e.g., sodium bicarbonate, sodium carbonate, and calcium carbonate) to strong acids and weak acids (e.g., citric acid) to strong bases (hydroxide solutions) results chemically in a neutral salt and water. The use of concentrated neutralizing agents (e.g., as sodium hydroxide pellets for acids) is *not* recommended.

Neutralization of spilled acids and bases requires the correct amount of neutralizing agent. Since the volume and molarity of acids and caustics differ, determination of the amount of neutralizer to be added to the spilled material is not an easy task. The reaction that occurs for neutralizing acids using a carbonate-bicarbonate mixture releases gaseous carbon dioxide, so that the lack of fizzing can signal the end point of a neutralization. Checking the resultant spill debris with pH paper will give more accurate results. The use of citric acid, on the other hand, to neutralize a caustic spill does not indicate complete neutralization by such foaming. The pH paper test is essential in this case.

Since neutralization is a strong exothermic reaction, the heat given off can cause splattering of the spilled material. For concentrated acids and caustics, as found in the majority of stockrooms, the addition of neutralizers at a fast rate will cause violent splatterings. Neutralization of these types of chemicals is necessarily a slow procedure. For very large spills (several gallons), neutralization cannot be considered a practical alternative.

Several manufacturers of spill kits for neutralizing acids and caustics have prepared prepackaged amounts of neutralizing agents. One manufacturer has formulated a neutralizing product (Neutrasorb™) that includes a color indicator for a neutralized acid spill. Although these kits are beneficial in that the rated capabilities are given, one must consider the economics of using them, as well as the bulk quantities needed for the potential of neutralizing large spills (Table 4.5).

Hydrofluoric acid deserves special mention because of its extreme corrositivity to human tissue. The acidic fluoride ion attacks skin quickly and without initial pain, but the resultant burns are hideously painful and slow to heal. A spill of hydrofluoric acid must be treated with a calcium-containing compound and soda ash to precipitate the fluoride ion as harmless calcium fluoride and render a neutral pH (26). Extra care must be taken to avoid skin contact with the acid, as well as inhalation of the hydrogen fluoride gas.

In summary, neutralization is accomplished by adding a neutralizing agent to a spill of a strong acid or base. A wet slurry still remains, which must be removed by absorption.

4.3.2.2 Absorption of Liquid Spills. Some liquid spills cannot be effectively treated before physically removed from a surface (e.g., organic solvents). The seemingly easiest way to remove a liquid spilled on the floor is to use a mop. The spill situation must be investigated, however, before such a method is employed. Mop heads made of cotton strands can be easily degraded by strong acids and oxidizing solutions, causing an

TABLE 4.5   Neutralizing Capabilities of Neutrasorb™ for Various Inorganic Acids

	Quantity[a] of Neutrasorb™[b]	
Acid	100 lbs	300 lbs
99% Glacial acetic acid	5.5	16.5
48% Hydrobromic acid	9.7	29.1
38% Hydrochloric acid	7.1	21.3
71% Nitric acid	5.6	16.8
72% Perchloric acid	7.1	21.3
87% Phosphoric acid	1.7	5.1
98% Sulfuric acid	2.4	7.2

[a]The figures in columns 2 and 3 indicate how many gallons of acid are neutralized by given quantities of Neutrasorb™.
[b]Neutrasorb™ is a trademark of the J. T. Baker Chemical Company.

additional mess. If water is added to dilute a spill of a strong acid so that a mop may be used, for instance, care must be taken to assess the corrosivity of the final mixture to prevent degradation of a metal mop bucket and wringer. Addition of water to a spill of concentrated sulfuric acid can be dangerous since the acid is water-reactive. Dilution of a spilled liquid with water to make it less hazardous will not work for all liquids, but only create a larger volume of liquid for removal from the floor.

Wringing a mop full of flammable liquids could cause a dangerous condition in the event that a spark is produced from the wringing mechanism. Because the recovered liquid is contaminated, it must be discarded properly. Solidification of the recovered liquid by using an absorbent is a step that should have been taken with the initial spill. A mop, wringer, and bucket should *not* be used on a spill of hazardous liquid, but may be acceptable and economical for nontoxic, noncorrosive, nonflammable, and nonreactive chemicals.

The use of an absorbent or gel to convert a liquid spill or slurry to a solid form and shoveling the residue into a waste container is the traditional method of absorption. Absorbents vary immensely in makeup, absorption capacity, inertness, and suitabilities for spills. Brugger gives a comprehensive listing of various vegetable (e.g., paper towels and sawdust), mineral (e.g., clay), and synthetic (e.g., polypropylene fibers) absorbents (27). A compatibility table of absorbents for various chemicals is listed in Ref. 27. Selection of a universal absorbent that will not break down or lose efficiency when in contact with all types of chemical is highly desirable.

The more inert traditional absorbents that can be used for treated and untreated spills of hazardous liquids are mineral in nature. These include exfoliated vermiculate, diatomaceous earth, sand, and granular clay. Since silicon is a major portion of the absorbent's chemical composition, these absorbents would react with hydrofluoric acid to produce noxious gases. Newer absorbents include a foamed amorphous silicate and treated polypropylene pads. All materials differ in cost-effectiveness and sorption capacity for various liquids. Hedberg (28) has done an initial study of the cost-effectiveness of several absorbents (Table 4.6).

Determination of how much absorbent is needed for predetermined volumes of spilled liquids is essential to plan to have enough absorbent on hand. Several manufacturers have packaged a predetermined quantity of foamed amorphous silicate in a porous polypropylene bag so that the absorbent capacity of the resulting "pillow" is rated at one gallon per

TABLE 4.6 Relative efficiencies of Adsorbents, Absorbents,[a] and Neutralizers

	Weight of Absorbent Before Immersion	Weight of Absorbent	Ratio Weight of Water Absorbed	Cost of Absorbent	Milliliters Absorbed	Milliliters Absorbed
	$t = 0$	$t = 10$ minutes	Weight Absorbent			$1.00
Polyolefin fiber	30 g	256	8.5	0.31	226	729
Cellulose fiber	30	191	6.4	0.06	161	2683
Vermiculite	30	87	2.9	0.037	57	1540
Calcium bentonite	30	58	1.9	0.009	28	3111
Foamed sand	30	271	9.0	0.096	241	2510
Starch polymer	30	1021	34.0	0.53	991	1870
Imbiber beads[b]	30	389	13.0	0.32	359	1122
Activated carbon[b]	30	49	1.6	0.32	19	59
Sodium bicarbonate[c]	30	—	—	0.03	30	1000
Citric acid[c]	30	—	—	0.16	71	444

[a]Protocol: 30 g of absorbent as placed in an 8½ × 8½ porous bag. Each bag was submerged in water for 5 minutes, allowed to drip for 5 minutes, and then weighted.

[b]Ethyl acetate was substituted with Imbiber Beads and activated carbon.

[c]Sodium bicarbonate and citric acid effectiveness was calculated with 12 N HCl and 6 N NaOH, respectively. Values in the table are corrected for absorption by bag.

Table courtesy of Lab Safety Supply Company, Janesville, Wisconsin.

bag. The speed of absorption by these pillows is quite rapid. The manufacturers' literature indicates that 98% of the rated capacity of a liquid is absorbed within 30 seconds. Such an arrangement also facilitates cleanup by eliminating the shoveling of the loose absorbent.

The volume and weight efficiency of an absorbent is important when considering waste disposal. The EPA has published legal regulations for the disposal of hazardous waste, including debris from hazardous materials spills (29). Absorbents that take up great volume will cause higher costs in disposal because of a greater number of disposal drums needed. Dense absorbents can cause a small-quantity generator to lose special privileges if a total weight of hazardous waste is exceeded (30).

The general procedure for using absorbents in cleaning liquid spills is (a) dike the spill (by surrounding it with a barrier of absorbent) (Figure 4.2), (b) add absorbent and mix, (c) repeat until a solid homogenous

FIGURE 4.2   Diking a spill with absorbent pillows to prevent the liquid from spreading. (Photograph courtesy of Lab Safety Supply Company, Janesville, Wisconsin.)

mixture results, (d) shovel the mixture into a chemically resistant disposal container, and (e) decontaminate the spill area. Note that step (d) requires knowledge of the corrosivity or flammability of the spilled liquid. Placement of an untreated absorbed acid into a metal drum will cause corrosion of the drum. Absorbents do not change the chemical hazards of a hazardous liquid. The advantage of absorption is speed of cleanup, which has inertness to the liquid as the main prerequisite.

4.3.3 Mercury Spills

Mercury is the only liquid metal listed in the Periodic Table. Mercury is extremely toxic through skin contact. Because there is no odor or warning properties, mercury vapors cause the greatest concern on spillage. Mercury is very mobile and as such can easily fill cracks and crevices and continue to vaporize until it is all gone. The traditional methods of cleaning mercury spills have been to sprinkle powdered sulfur over the contaminated surface and precipitate mercuric sulfide. Recently, commercially prepared powders have been designed to amalgamate the mercury on contact. Foam collectors have been used to recover small droplets of mercury. Specially impregnated granular activated carbon has been used to cover suspected areas of mercury contamination so as to "adsorb" the vapors without contaminating room air. In a chemical stockroom that stores kilogram quantities as well as scientific apparatus such as barometers and manometers, purchase of a specially designed vacuum cleaner is recommended. Figure 4.3 shows such a vacuum cleaner that incorporates a device for separation of mercury from the rest of the spill debris by centrifugal action. The mercury is then deposited in a separate reservoir. The vacuum cleaner is equipped with charcoal filters to prevent the release of mercury vapors. Since mercury vapors cannot be detected by the senses, completeness of decontamination after a mercury spill must be measured by using either impregnated mercury indicator paper (which changes color in the presence of mercury) or a special mercury monitor.

4.3.4 Spill Response Centers

Once the appropriate response equipment has been chosen, a readily accessible area must be designated as a spill response center. In this area,

FIGURE 4.3 A vacuum cleaner designed for recovering spilled mercury. The mercury separator on the front operates in separating mercury from other debris by centrifugal action. (Photograph courtesy of the Marketing Communications Department, Hako Minuteman, Addison, Illinois.)

the necessary absorbents, neutralizers, protective gear, and other paraphernalia can be assembled and stored until use. It is important to establish an inventory–inspection system so that depleted stocks may be replenished to have sufficient amounts on hand for future spills. It is advisable to have a mobile cart or truck that will promptly transfer the necessary materials to the spill scene. One manufacturer has designed such a response unit for small spills (Figure 4.4). The spill response team will need to have access to spill cleanup procedures. This area must have instructions and chemical hazard data files available to trained personnel.

FIGURE 4.4 A mobile spill response cart, complete with spill handling materials and protective clothing kits. (Photograph courtesy of Lab Safety Supply Company, Janesville, Wisconsin.)

4.4 SELECTING PERSONAL PROTECTIVE EQUIPMENT

Since hazardous chemical exposure resulting from a spill scene can cause damage to human health or life, proper selection and use of protective equipment is an absolute requirement. Since the routes of exposure include inhalation, eye and skin contact, appropriate respiratory protection, clothing, gloves, and eye protection must be determined in advance. The spill response team must have familiarity with the equipment and its use. Finally, inspection, maintenance, and repair must be kept current to avoid unnecessary accidents.

4.4.1 Respiratory Protection

Inhalation of toxic gases, mists, vapors, particulates, or any combinations thereof are possibilities when dealing with chemical spills. Solid spills will produce dusts that require the use of a mechanically filtering mask. Spills of harmless solids may produce effects that require only a NIOSH-approved disposable dust mask. Spills of toxic solid chemicals, on the other hand, require greater protection. A NIOSH-certified HEPA filtered gas mask or cartridge respirator will serve well for those toxic dusts for which the respirator is recommended. Some dusts (e.g., carcinogens and cyanides) pose chronic hazards or are extremely lethal and may require the use of a NIOSH-certified pressure demand self-contained breathing apparatus (SCBA).

Liquid spills are far more complex when determining the proper respiratory protection. Liquids can produce organic vapors, acid gases, mists, or combinations thereof that may or may not be effectively stopped by an air-purifying respirator. The concentration and nature of these vapors or gases is critical for selecting the correct respirator. The data in Table 4.4 indicate that the saturated vapor concentration can be many times over the concentration level determined to be immediately dangerous to life or health. With this in mind, the spill scene can be best considered an emergency that requires the use of a NIOSH-certified pressure demand self-contained breathing apparatus.

The possession and use of respirators is a necessary cause for compliance with the minimally acceptable respirator program specified by OSHA (31). Training and fitting must occur *before* personnel respond to the spill scene. Manufacturers' instructions show how to operate and fit their respirators. Pending OSHA legislation will require the fitting of respirators while worn in a test atmosphere. A valuable guide to the operation of a SCBA is *Self Contained Breathing Apparatus, First Edition* by IFSTA (32). Inspection, maintenance, and repair of respirators must be done regularly. Additional helpful information is available on selecting respirators (33,34). Approved and accepted respirators can be determined by consulting the *NIOSH Certified Equipment List* (35). Personnel who are selected for the spill response team must have medical examinations to determine their physical capability to wear and use a specified respirator.

4.4.2   Protective Clothing

Splashes and contact with chemicals can occur in a spill situation. The use of protective clothing to provide a barrier to this contact is essential. Clothing that holds out liquid and particulate matter is desirable. The style of clothing should be appropriate to the spill. Aprons, disposable coveralls, and totally encapsulated chemical splash suits may be demanded by the size of the spill, toxic and corrosive nature of the spilled chemical, and the type of spill response method employed.

Disposable clothing that is impenetrable to liquid and particulates fits the needs of most chemical storeroom spills. Jumpsuits and coveralls made of polyethylene-coated Tyvek® (36) are lightweight, are easily stored, and have good protective capabilities (Table 4.7). Additional features such as elastic wristband, attached hood, and boots help to provide comprehensive covering for the body.

4.4.3   Gloves

Gloves protect the hands from exposure to chemicals. The hands come into contact with the spilled material very readily, which may be corrosive to skin tissue or absorbed through the skin in small yet toxic doses. Thus proper selection of protective gloves that prevent skin contact with the chemical is of prime importance. Gloves must be selected on the basis of resistance to (a) chemical degradation, (b) permeation, and (c) penetration.

There is no glove that is free from degradation by all chemicals. Contact with certain chemicals will cause softening, swelling, embrittling, cracking, and dissolving of the glove material so that the barrier properties of the glove are lessened and skin contact imminent. Consulting the glove manufacturer's literature and even field-testing glove materials will guide the selection of the proper glove for use with a specific group of chemicals. Choosing an appropriate glove for protection from inorganic acids and caustics is much easier than choosing a glove for the variety of organic solvents. One may need several pairs of gloves made from different materials to effectively resist degradation from the wide variety of organic solvents.

''Permeation'' is defined as the molecular diffusion of liquids in contact with the exterior surface of a glove through the glove material so that the

TABLE 4.7   Barrier Performance of Tyvek® to Hazardous Materials

Hazard	Tyvek®		Polylaminated Tyvek®		Saranex-Coated Tyvek®	
65% Oleum	—	(—)	<1	(Na)	37	(Na)
20% Oleum	—	(—)	120	(Na)	—	(—)
98% H_2SO_4	<5	(3000)	>480	(ND)	>480	(ND)
96% H_2SO_4 (65°C)	—	(—)	>120	(ND)	330	(Na)
90% H_2SO_4	<5	(2300)	—	(—)	—	(—)
50% H_2SO_4	6	(270)	—	(—)	—	(—)
16% H_2SO_4	30	(55)	—	(—)	—	(—)
90% HNO_3	—	(—)	—	(—)	107	(Na)
70% HNO_3	—	(—)	50	(Na)	>48 hours	(ND)
37% HCl	—	(—)	35	(Na)	>48 hours	(ND)
100% HCN	—	(—)	60	(111)	—	(—)
10% HCN	—	(—)	>120	(ND)	—	(—)
TiCl	—	(—)	—	(—)	(>1000)	(ND)
45% NaCN (70°C)	—	(—)	240	(6)	—	(—)
10% NaCN (60°C)	—	(—)	360	(<1)	—	(—)
Cl_2 (20 ppm)	—	(—)	>480	(ND)	>480	(ND)
Sodium chromate Cleaning solution	80	(34)	>480	(ND)	>480	(ND)
40% NaOH	10	(38)	>480	(ND)	>480	(ND)
Mineral spirits	—	(—)	<5	(7)	>10	(<0.2)
o-Toluidine	—	(—)	<5	(1)	>120	(<0.03)
Oxydianiline	<90	(Na)	270	(NA)	—	(—)
Toluene	—	(—)	<5	(165)	<5	(20)
Chloroform	—	(—)	<1	(348)	<1	(201)
Trichlorobenzene	—	(—)	<15	(5)	15–60	(0.04–0.1)
Tetraalkyl lead	—	(—)	<30	(8.36)	60	(0.079)
PCB	—	(—)	<60	(0.0002)	60–120	(0.0002)
57% Methyl parathion	—	(—)	15	(0.09)	120–180	(0.01)
10% Methyl parathion	<5	(45)	30–45	(0.2)	>240	(<0.002)
Cresols	—	(—)	40–60	(0.4)	>120	(<0.14)
Acrylonitrile	—	(—)	5	(0.0006)	23	(0.0013)

®Breakthrough time in minutes is indicated by the first number for each type of Tyvek® material. Permeation rates in micrograms per minute per square centimeter are presented in parentheses. Abbreviations: NA, not measured; ND, none detected.
Table courtesy of the Du Pont Company of which Tyvek® is a registered trademark.

liquid appears inside the glove. The amount of time for permeation to occur is called the *breakthrough time*. The rate of permeation may be rapid or slow, so that minute or sizable quantities of liquid appear inside the glove. There is not much current data on permeation of gloves because of the immensity of testing needed. The breakthrough time and permeation rate depend on the glove material and its thickness and the particular solvent. Some initial work has been done by the Edmont Wilson Glove Company on their line of neoprene, nitrile, polyvinyl alcohol, and polyvinyl chloride gloves. Viton® (37) and butyl rubber gloves also offer varying resistance to permeation and excellent resistance to chemical degradation. The reader must consult manufacturer literature before making the selection.

Penetration of a glove material occurs because of small imperfections or lesions in the material itself. Penetration most commonly occurs at the seams of the gloves. In cases where critical protection of the skin tissue is necessary, the practice of submerging an inflated glove in water to detect leaks is recommended. Commercially available glove inflators, such as those used for lineworker's gloves, may be applicable in this type of testing.

4.4.4   Eye Protection

American National Standards Institute Standard Z87.1 (38) requires that chemical splash goggles be used whenever there is a danger of chemical splash. Since industrial hygienists and physicians recommend that eyes that have been contaminated with chemicals be flushed with water for 15 minutes, it would behoove personnel to wear protective goggles. Such goggles have indirect ventilation ports that direct any liquid away from the eyes. Face shields may be necessary in some instances where the face may receive chemical splash. The spill situation should be imagined creatively, so that the details of providing protective equipment are not postponed until the last moment.

4.5   SELECTING AND TRAINING THE SPILL RESPONSE TEAM

The spill response team is the crux of the entire effort to effect spill cleanup efficiently and safely. Selection of the right personnel who will be designated on-call to respond to a spill without notice cannot be overemphasized. Simply selecting personnel because they show up for work without regard for other factors will result in an unfavorable decision. Personality traits of prudence and caution and ability to respond to emergencies effectively must be weighed. Since a team effort is required, inclinations to comradery and working with others is essential. Clear thinking under pressure, the ability to follow instructions. and persistence are necessary. Members of the response team should not be overly sensitive to chemical exposure. It is important that screening be done to establish the medical history of candidates for the response team. No less than two persons should comprise an actual spill response unit. Since dangerous conditions exist, communication devices (if allowed by the circumstances) should be employed. In the final analysis, selection of the spill response team must be done through a careful decision matrix.

Training the team must include familiarization with the types and hazards of storeroom chemicals, spill response methods, the designated spill response files, and protective equipment. Hands-on experience during drills and mock spills must be provided. Review of safety precautions and disposal methods must take place. The cleanup equipment must be used and understood prior to the emergency spill. Training exercises may also include lectures, the use of films, and demonstrations. The application of first aid procedures should be studied and reviewed. Assessment of training programs of other institutions can cause a beneficial information exchange to broaden the training base (39,40).

4.6 CONCLUSION

Chemical spills associated with the storeroom may be large or small, hazardous or nonhazardous. Many spills can be avoided by using proper management controls. For those that do occur, an informed plan of action, properly selected equipment, and a trained response team can turn a potential disaster into a routine matter. A safe and effective storeroom cannot be established until the issue of spill response is resolved.

REFERENCES

1. J. S. Robinson (Ed.), *Hazardous Chemical Spill Clean Up,* Noyes Data, 1979.
2. G. Bennett, F. Feates, and I. Wilder (Eds.), *Hazardous Materials Spills Handbook,* McGraw-Hill, New York, 1982.

3. A. Smith, *Managing Hazardous Substance Accidents,* McGraw-Hill, New York, 1981.

4. D. Hedberg, "Spill Control in the Chemical Storeroom," presented at 187th National ACS Meeting, Kansas City, September 1982.

5. W. T. Niggel, "Leakage and Overpressurization of 55 Gallon Drums," in *Proceedings of 1982 Hazardous Materials Spills Conference,* Milwaukee, WI, April 1982, pp. 482–486.

6. National Research Council, *Prudent Practices for Handling Hazardous Chemicals in Laboratories,* National Academy Press, Washington, DC, 1981.

7. M. Lee, "Development of a Local Hazardous Materials Management System," in *Proceedings of 1982 Hazardous Materials Spill Conference,* Milwaukee, WI, April 1982, pp. 147–150.

8. U.S. Coast Guard, *CHRIS Hazardous Chemical Data,* Government Printing Office, Commandant Instruction M16465.12, October 1978.

9. *Handling Chemicals Safely,* Dutch Association of Safety Experts/Dutch Chemical Industry Association/Dutch Safety Institute, Netherlands, 1980.

10. *Toxic and Hazardous Industrial Chemicals Safety Manual,* International Technical Information Institute, Tokyo, Japan, 1982.

11. *NIOSH–OSHA Pocket Guide to Chemical Hazards,* U.S. Government Printing Office, NIOSH Publication 78-210.

12. N. Sax, *Dangerous Properties of Industrial Materials,* 5th ed., Van Nostrand-Reinhold, New York, 1979.

13. *1980 Emergency Response Guidebook for Hazardous Material,* U.S. Department of Transportation, printed by Labelmaster, Chicago, 1980.

14. General Electric Company, *Material Safety Data Sheets,* Schenectady, NY, 1980.

15. U.S. EPA, *Hazardous Material Spill Monitoring: Safety Handbook and Chemical Hazard Guide,* Part A, National Technical Information Service Publication No. PB 295853, January 1979.

16. M. Lefevre, *First Aid for Chemical Accidents,* Academic Press, New York, 1980.

17. Code of Federal Regulations, Title 40, Sections 261–265.

18. ACGIH "TLVs—Threshold Limit Values for Chemical Substances and Physical Agents in the Work Environment with Intended Changes for 1982," ACGIH, Cincinnati, OH, 1982.

19. These must be calculated. For a discussion of evaporation rates, see D. Machay's et al. article, "Calculation of the Evaporation Rate of Volatile Liquids" in *Proceedings of 1980 National Conference on Control of Hazardous Material Spills,* Louisville, KY, May 1980.

20. D. Pipitone and D. Hedberg, "Safe Chemical Storage: A Pound of Prevention is Worth a Ton of Trouble," *J. Chem. Ed.* (May 1982).

21. M. Pitt, "A Vapour Hazard Index for Volatile Chemicals," *Chem. Ind.,* 804–806, October 16, 1982.

22. Reference 6, p. 235.

23. J. Meidl, *Flammable Hazardous Materials,* 2nd ed., Glencoe Press, Encino, CA, 1978, p. 168.

24. Reference 10, p. 426.

25. D. Hedberg, "Clean Up of Chemical Spills in Labs," *Natl. Safety News* (March 1981).

26. Reference 10, p. 276.

27. J. Brugger, "Selection, Effectiveness, Handling, and Regeneration of Sorbents in the Clean Up of Hazardous Material Spills," in *Proceedings of 1980 National Conference on Control of Hazardous Chemical Spills,* Louisville, KY, May 1980, pp. 92–98.

28. D. Hedberg "Chemical Spills in the Storeroom," in *Symposium on Prudent Practices for the Safe Storage of Chemicals,* Division of Chemical Health and Safety, American Chemical Society, 187th National Meeting, Kansas City, MO Sept. 1982.

29. Federal Register, Monday, May 19, 1980, Code of Federal Regulations, Title 40, Section 261.33, Paragraphs (e) and (f).

30. Code of Federal Regulations, Title 40, Section 261.5 (c) (4).

31. Code of Federal Regulations, Title 29, Section 1910.134.

32. IFSTA, *Self Contained Breathing Apparatus,* 1st ed., Fire Protection Publications, Oklahoma State University, Stillwater, OK, 1982.

33. ANSI Standard Z88.2-1981, "Practices for Respiratory Protection, ' American National Standards Institute, 1430 Broadway, New York, NY 10018.

34. John Pritchard, *A Guide to Industrial Respiratory Protection,* DHEW Publication No. 76-189, NIOSH Cincinnati, OH, 1976.

35. *NIOSH Certified Equipment List—1981,* DHHS Publication No. 81-44, NIOSH, Cincinnati, OH, 1981.

36. Tyvek® is a registered trademark of the Du Pont Company.

37. Viton® is a registered trademark of the Du Pont Company.

38. ANSI Standard Z87-1, 1979, "Practice for Occupational and Educational Eye and Face Protection," American National Standards Institute 1430 Broadway, New York, NY 10018.

39. A. Moreisi, "Organizing an Environmental Response Squad to Meet Superfund and RCRA Corporate Responsitility," in *Proceedings 1982 Hazardous Materials Spill Conference,* Milwaukee, WI, April 1982, pp. 241–246.

40. G. Tompkins and R. Garton, "Hands on Training for Industrial Emergency Response Teams," in *Proceedings of 1982 Hazardous Material Spills Conference,* Milwaukee, WI, April 1982, pp. 497–499.

CHAPTER 5

USE AND SELECTION OF COMPUTERS FOR CHEMICAL TRACKING SYSTEMS

ALLEN G. MACENSKI
Hughes Aircraft Company, El Segundo, California

5.1   INTRODUCTION

Chemicals occur in almost limitless (and ever-increasing) varieties. For this reason, general precautions for handling almost all chemicals are needed, rather than specific guidelines for each chemical. Otherwise, laboratory work would be handicapped, practically and economically, by attempts to adhere to a labyrinth of separate specific guidelines. What is more likely in this instance is that the laboratory worker will simply ignore the entire complex set of guidelines, and, consequently, be exposed to excessive hazard.

However, each chemical poses different hazards. Data on the extent and type of hazard are tremendously difficult to compile, to say the least. Careful attention must be paid to the appropriateness of the work to be done in relation to the properties of the chemical(s), the adequacy of the physical facilities available, and the personnel involved. Once these are established, it is the role of the safety coordinator and the representative group to assist in the development of adequate guidelines for operations. But the starting point for this planning effort is related to the properties of the chemical(s). Since the vastness of chemical formulation is overbearing, aids in data gathering and collection are essential for today's health and safety professional. The computer is now becoming a required tool for the efficient mastery of chemical safety information.

The needs are to (a) evaluate work place health and safety factors, (b) identify possible cause/effect relationships to health variables, and (c) monitor the toxic effects of the company's material on employee health. The commitment is to create a viable storehouse of historical and current data regarding workplace conditions and materials and each employee's work history as exposed to the stress. This necessitates collection of uniform, pertinent data from a number of disparate sources and then storage and maintenance of those data in an accessible manner. A viable system must have the ability to retrieve information on various relationships and cross-tabulate these answers and must have detailed information on the toxic substances handled or manufactured—and details on the effects of those substances.

The user must have the ability to obtain answers to ad hoc questions and be able to combine and analyze all the data in the various subsystems in a unified manner and to assess the impact of proposed legislation and regulations on a "what if" basis. And to do all this, the potential user must go back to the starting point—collecting uniform, pertinent data from a number of disparate sources.

The following list is minimal, but it should suffice to show how any in-house system, manual or computerized, can potentially strain corporate resources to the limit:

Materials and Substances

Material Safety Data Sheet (MSDS) information
Locations
Exposure risk factor
Use
Synonyms

Industrial Hygiene

Personal and area samples
Job description
Employee work location
Level of exposure data

Safety

Documentation of training
Accident Records
Accident frequency rates

Medical

Preemployment medical examination information
Employee-provided medical history
Accident and illness record
Medical tests related to specific employees
Workplace hazardous exposure data

Toxicology

Material toxicologic codes

Epidemiology

Medical records

Illness and death statistics

Exposure information

Demographic employee profile

5.2   DIFFERENT COMPUTERS FOR DIFFERENT TASKS

Unless you have been sequestered in your office with the shades drawn and the phone unplugged during the past few years, you already know that computers have escaped from the computer room and are invading the secretarial pool and the executive suite. They travel in briefcases and speak English. In short, they are becoming ubiquitous in the world of American executives as paper clips and perquisites.

Business computers are now advertised on television between innings of baseball games. Business computers are touted in the magazines *Time* and *TV Guide*. They may be ordered by mail or carried home in shopping bags from any of a hundred or more catalog houses or department stores—and, of course, from any of the growing number of computer boutiques sprouting up in suburban shopping malls and on downtown thoroughfares. Electronic gadgetry is having a profound effect on the business lives of an increasing number of us as computers—once considered to be the quint-essential banes of our existence—have become small and friendly enough to go just about anywhere. Many such "personal business computers" cost less than $4000.00 and are capable of performing technological feats that not so long ago required a computer four times the size and eight times the cost.

Today, there are increasing number of choices—in the machines them-selves, in the software needed to run the machines, and in the places from which to buy. The resulting confusion can be formidable for an executive who wants to invest capital in computers without investing a lot of time sorting through the business–computer jungle. To make matters worse, there is a massive information gap between computer manufacturers and buyers, particularly among corporate health and safety professionals and management, who all too frequently consider themselves too busy to take the time to learn about this vital technology.

Perhaps the most surprising revelation by computer makers recently is that microcomputers—small, desk-top models that need not necessarily be tied into a larger system—are not just for microbusinesses. Among other things, the attraction of microcomputers is that they offer flexibility in computing. The "micros" can serve as efficient stand-alone appliances that can enable safety directors to easily plot accident trends, answer some "what if" questions, or merely zip memos electronically to colleagues or subordinates; at the same time, they can plug into a larger mainframe computer system in order to gain quick access to additional data files and more sophisticated programs.

Before the uninformed professional ventures into the computer world, understanding of the terminology is essential.

5.2.1 Understanding Computer Buzzwords

Each field of endeavor has its own "buzzwords," which seem to exist only to confuse the uninitiated and to perpetuate the mystique. Few fields of endeavor can surpass data processing for an abundance of buzzwords. Before an intelligent discussion can begin, some basic terms must be defined.

First, consider the term "computer." Overused, it becomes a catchall for an incredible variety of appliances and electronic contraptions. Today, computers can be found everywhere, from automobiles to microwave ovens and from video games to massive mainframes used to control entire manufacturing plants. The first step toward an adequate understanding of electronic data processing (EDP) is an understanding of the generic classes of computers or "hardware," which are shown in Table 5.1.

The one statement that can be made about the differences between the various kinds of computer is that there are no absolutes. Rather, wide gray zones exist between classes of computers, and reasonably informed experts can debate whether a computer is really a "home computer" or a microcomputer. Similarly, many of the so-called special-purpose word processors are, in fact, microcomputers that can be used for other purposes.

There are a few guidelines, though, for differentiating between classes. Although it may be relatively devious as to what constitutes a mainframe, the dividing line between mainframe and a minicomputer is more difficult to see. Some minicomputers use the same internal designs as mainframes,

TABLE 5.1   Generic Classes of Computers

	Buzzword	English Equivalent
Large	Mainframes	Large-scale equipment, manufactured by firms such as IBM, Control Data, Cray, and others; these are the business machines that typically consist of a whole room full of equipment
	Minicomputers	Medium-scale equipment, such as the IBM System 34, the BASIC 4 computer, and the Hewlett-Packard 3000; these machines are generally based on a scaled-down mainframe design
Small	Microcomputers	Small-scale implementations of a computer—the Radio Shack TRS 80 is a good example; probably the fastest-growing segment of the computer field
	Personal computers	The smallest of the microcomputers, although there is no clear dividing line between the two categories; the Apple Computer is a good example
	Word processors	Special-purpose microcomputers, but many products have as much capability as some of their general-purpose relatives

so perhaps a price guideline is best to employ. Suffice it to say that if it costs less than $300,000 today, it most probably is a minicomputer.

Drawing the line between minicomputers and microcomputers is equally difficult. A rough rule of thumb is that if the computer fits on top of a normal-sized office conference table, it probably is a microcomputer. If it has its own cabinet and is desk-sized, chances are it is a minicomputer. Another difference is the storage capability: minicomputers generally include a "hard disk" for storage that has removable "disk packs"; microcomputers generally don't include this hardware feature.

Computers can also be distinguished by another factor relating to their internal communications line (wires), which are used to communicate with the system memory and other internal components. These internal communications channels carry digital signals, so called because they consist of electrical impulses that represent either zeros or ones (0's or 1's) (bits). Each computer reads, works, and outputs these signals in uniform-sized batches (bytes) consisting of a certain number of 0s and 1s, the size of which is known as a "word length."

In the first computers, a length of four bits was used, which meant that each batch had four characters, each of which could be either a 0 or a 1. Gradually, over the years, a sophistication increased, so did the word length. In mainframe computers, a length of 32 bits is commonly used. Minicomputers generally used a word length of 16 and microcomputers could be easily defined as those computers that used an 8-bit length.

However, just as soon as everyone in the computer business thought they had the definitions down pat, someone brought out a 16-bit microcomputer, and some of the minicomputers started using 32 bits. So much for trying to keep things simple.

5.2.2. The Myth About Minicomputers

Health and safety professionals of today are generally interested in word processors, home computers, microcomputers, and—occasionally—a minicomputer. What may come as a surprising revelation is that one of these machines is intrinsically any better than any other; each excels in a certain environment. The key to selecting the right machine is an understanding of what it is best suited for, as well as a good understanding of the production demands that will be placed on it. The classic example of market misinformation is the continuing debate between minicomputer salespersons and microcomputer dealers.

Minicomputers are thought by some to be superior to microcomputers for any business application. Holders of this opinion use the argument that microcomputers are unreliable, unprofessional, and built more for use as a hobby than for business. Nothing could be further from the truth. Whereas the smaller models are obviously unsuitable, most microcomputers are generally built with the same components used in smaller minicomputers. The two can generally perform the same tasks with almost the same operator speed. In many cases they use the same printers, the same terminals [cathode ray tubes (CRTs)] and employ the same storage devices (floppy disks or hard disks).

5.2.3 A Glossary of Computer Components

There are a number of fundamental computer terms and concepts with which health and safety professionals should be familiar. The following list details the various components that comprise a computer system, and why they are important.

1.   *Input devices* translate information into electrical signals that are intelligible to the computer. Many different kinds of device exist, the most common of which is a typewriterlike keyboard. There are also some sophisticated devices, including document readers that can scan entire pages of material and convert them into computer input, light pens that allow operators to draw images on the computer screen, and even speech recognition devices that allow users to enter information orally into computers.

2.   *Storage devices* allow users to store information when the computer is turned off. There are many kinds of storage devices, but most of the systems that safety engineers and industrial hygienists will use employ floppy disks. A strange-sounding name for a very popular and useful storage method, floppy disks are available in three sizes—8, 5¼, and 3½ inches. They consist of a square plastic jacket, in which a thin plastic disk is free to rotate. A wide slit on one side of the disk exposes a gray–brown magnetic surface, onto which magnetic impulses are recorded by the disk drives into which the disks are inserted when used. The name "floppy disk" is derived from their relative lack of physical rigidity when compared to hard disks, another storage medium used in larger computers.

3.   *A computer's internal memory* is known as *random access memory* (RAM). Memory is measured by the number of bytes of information that can be stored in it. Typically a computer has at least 48K (48,000 bytes) of RAM, although some systems have as much as 500K (500,000 bytes) or more.

The fact that one computer has more RAM than another does not necessarily mean that it is any more powerful or easier to use, as some salespeople would like to imply. In the computer, RAM consists of one or more small black plastic cases with eight or more connector pins on either side, measuring about 1/4 inch wide and about 1 inch long. They are sometimes called "chips," because of the small silicon wafers that constitute the memory components.

4.   *The central processing unit* (CPU) in early computers used to consist of a gigantic rack full of equipment, with flashing lights and humming fans. Now computers normally consist of one rather unimpressive-looking plastic case about 1 inch wide and 3 inches long, with a number of leads on either side. However, inside this wonder chip are the

equivalent of up to 64,000 transistors, diodes, resistors capacitors, and other electronic components. Thanks in large part to the space program, a complete computer central processing unit that formally occupied a room and cost (in 1957 dollars) over $500,000 can be bought over the counter at an electronic store for $500 today. There are several different kinds of CPU chip. The type of chip is more important to the user than the brand of computer because it will determine what software is available. The byte length used in most small computers is either 8 or 16 bits. Eight-bit computers have been around for several years, using several chips, among them the 8080, 8085, Z-80, Z-80A, and 6502. Sixteen-bit chips are relatively new on the market.

5.   *Output devices* are, for general purposes, printers that come in all sizes, speeds, and shapes, from those that are no more than converted IBM Selectric typewriters to sophisticated line printers capable of printing up to 1000 lines per minute.

5.3   COMPUTER SOFTWARE

This brief introduction to computer technology would be only half complete without a discussion of software—the programs that give the computer its personality and determine in large part its capability.

Computers are, by their construction, very complex machines with no intellect. Software provides the machine with the intelligence necessary to function as a computer. But there is more to software than meets the eye of the uninitiated.

At least four levels of software exist: the machine "monitor program," the operating system, the language, and the applications program. All of these levels of software are present in a machine at any time and play a vital role in the functioning of the machine. The monitor program is an elementary program level that is stored in a special kind of memory called *read-only memory* (ROM). When the computer is first turned on, this program is invoked to do the basic things such as set the CPU running correctly and perform some initial start-up functions that are not really of much interest to other than a programmer.

The operating system, on the other hand, is of vital concern to the user. An operating system is a complex piece of software that is usually the first thing that the computer "loads" into memory when started. As its name somewhat implies, the operating system provides the computer with its basic operating personality and sets certain protocols and rules that must be used in dealing with the system.

The next level of software is the language, which in the case of most mini- and microcomputer applications is usually BASIC or COBOL. These languages provide programmers with a means of communicating with the computer. The results of these communications are lists of instructions that are stored as applications programs and called forth from the storage devices when needed.

When a person sits down at a computer, the following happens: a program (an application program) is run; instructions are communicated to the computer by means of a language; and the language interacts with an operating system that deals with the computer to give it a certain personality. Sounds complex? It is, and it all takes place in a split second without the user even knowing (or in some cases caring) about it. However, knowledge is power, and the astute health and safety professional should understand the basics of software in order to make intelligent decisions concerning the uses and limitations of computers in their practices.

5.3.1   Some Hard Facts About Software

One of the most frustrating parts of launching into computers is finding the right kind of computer software—the instructions that tell your computer how to perform various tasks, from inventory to chemicals to dose rates per employee to word processing. If there is anything more baffling than sorting through the ads and literature to choose the right computer, it is deciding which software will best suit your needs.

There is one school of thought that says you should shop for the software *before* you shop for the hardware—the physical machine and all its accessories. The idea makes sense. After all, you probably wouldn't buy a video tape recorder if you know that there were only three taped programs that could be played on it—unless, of course, you intended to produce your own. The same is true with computers: unless you can write programs yourself, you would be foolish to buy a piece of equipment that couldn't do what you wanted it to do.

By the way, don't scoff at the notion of writing your own programs. If you're so inclined, it's not all out of the question. If you learn the basics of BASIC—which stands for "beginners' all-purpose symbolic instruction code," the computer language in which many popular computer programs are written—you can write several programs yourself.

On the other hand, some software programs are already available for use on the mainframe computers. In addition, Digital Equipment Corporation, Marlboro, Ma., has a complete package, both hardware and software, available called "DEC health" (medical and environmental informaton system) that will allow both medical, and health and safety personnel to compile demographic exposure, medical, and mortality data for their organization.

5.3.2 Time Sharing

The DTSS Time Sharing Service is a key example of the comprehensive software system for interactive, remote access computing. The DTSS was developed at Darmouth College in a project begun in 1964 by Professors John Kemeny and Thomas Kurtz. This early system serving Dartmouth students and faculty was the first to make time-sharing computing available to nonprofessionals. Other educational institutions plus a limited number of businesses soon were permitted to access DTSS by remote terminal. When computer service and commercial companies began to apply for licenses to operate DTSS software on their own computers, nonprofit Dartmouth College decided to divest itself of what was becoming a major business enterprise.

Known for its "user-friendly" design, DTSS has simple, English language commands and clear, informative diagnostics; DTSS can serve novice and expert users alike.

Another general-purpose-type database management system is "Systems 2000" (S2K), which is vended by M. R. I. Systems Corporation in Austin, Texas. Basic Systems 2000 provides the user with the ability to define new databases, modify existing databases, and retrieve and update the values in a database. It also provides the capability of saving the database on tape for backup and recovery of damaged databases.

"Procedure language interface" enables users to manipulate data in a System 2000 database from either a COBOL or FORTRAN program. This allows the user to access two or more databases in more complex applications. Immediate access provides the user-oriented, English-like,

S2K Natural Language for retrieved and update of the database. The S2K system also provides statistical functions (SUM, COUNT, AVERAGE, MAXIMUM, MINIMUM, STANDARD DEVIATION) and a simple reporting technique that features sorting and data selection capabilities.

Report writing enables the user to prepare reports using the S2K Natural Language. Up to 100 reports may be generated from a single pass of the database index files and may include the following: break points on any database element or group, page heading, page footings, or page ejection and the accumulation of subtotal or grand-total dynamically through conditional statements.

The Multi-User permits communication between two or more regions using a single copy of S2K. Batch and on-line users may ACCESS the same database simultaneously.

5.3.3   Dialing for Data

Another reason why top managers of health and safety programs are jumping onto the microcomputer bandwagon is the rich code of "on-line" information resources that computers open up. By simply plugging a desktop computer into a telephone line—which requires an inexpensive device called a modem—a computer user can have access to any of nearly 1000 on-line data banks, some of which have direct application for occupational health and safety. Users pay anywhere from $5 to $300 per computing hour for using these services. The data banks open up a brave new world of information. For example, there is Medline, a database available from the National Library of Medicine, part of the federal government's National Institutes of Health. Medline lists some 4 million references to articles published over the past 15 years from thousands of medical journals. To use Medline, one types key words into their computer; Medline responds with a complete listing of entries under that key word.

At the 183rd American Chemical Society's National Meeting (Division of Chemicals Information), Dr. Po-Yung Lu of the Oak Ridge National Laboratory in Tennessee and Carol B. Haberman of the National Library of Medicine described the national Toxicology Data Bank (TDB). The TDB is another on-line interactive data retrieval file for chemical, toxicological, and environmental effects information of potentially hazardous substances. It contains 2800 complete records with an approximate 900 additional compounds in various states of preparation. The TDB is updated quarterly and a part of MEDLARS, National Library of Medicine (File—TDB).

The list of data banks continues: RTECS, a service of the National Institute for Occupational Safety and Health in Cincinnati, lists information on 47,000 toxic chemicals; Bioethics line, from Georgetown University, offers citations on subjects such as human experimentation; and Smithsonian Science Information Exchange provides references to 160,000 on-going and completed scientific research projects. There are even data banks of data banks. One company, "Information on Demand," maintains access to more than 200 data banks that can yield facts on everything from the weather to usage levels of strategic metals. One starting point for adding data to one's resources should be the U.S. Commerce Department's Bureau of the Census, which maintains a list of telephone contacts for data users. Many states and universities have data centers that can guide the user toward the information needed.

5.4 FINDING THE RIGHT COMPUTER

How do you find the "right" computer? It takes equal parts of hard work and luck. Among other things, you should understand that the "right" computer is one that will fit your needs today as well as a few years from now. That requires expandability—the ability to add storage capacity, telecommunication, printing devices, and more powerful operating software as your business demands it.

One of the growing number of independent computer stores may be a good place to start your search. For one thing, such stores sell a number of competing brands of computers, so salespeople there can direct you to the one best for you.

Another word of advice: don't be afraid to ask questions. Or, as another sage put it, the only dumb questions are the ones you don't ask. Here are some other questions from experts to help guide you through the computer marketplace:

1. Conduct a thorough analysis of your automation needs. Build a matrix to determine who communicates with whom. Design a questionnaire to solicit employee input from your own and other departments. Make exhaustive lists, drawings, and charts—whatever you need to get a firm grasp of your needs.

2.   Conduct a lease–purchase evaluation. In many cases, the advantage of one over the other may be overwhelming. Generally, there is no reason to buy equipment unless the pay-off period is very short—a year or two. Consider the length of the lease, too. It's unwise to get locked into a 5-year lease when you may outgrow the system in 3 years.

3.   Don't be sidetracked by the notion of obsolescence. If you want to get technical about it, everything presently on the market is obsolete because several dozen manufacturers have things on the drawing boards that are better and more powerful than what's now available. In a sense, what you buy will always be obsolete—but only from a technological point of view. Your equipment is only obsolete when it fails to serve your needs.

4.   Consider documentation, training, and servicing. The finest machine does you little good if the operating instructions are incomprehensible. Many computer manufacturers provide extensive training for employees. Servicing is another factor that varies widely among manufacturers, but few computer buyers consider it, although it can cost plenty when a system malfunctions. Find out how big the servicing network is, how close the nearest outlet is to your office, and what the average response time is. Check with others who have purchased similar equipment to see how satisfied they are with service.

5.4.1   Eight Things Your Computer Won't Do

1.   *A computer won't save you money.* In fact, it may cost you a bit because of the often overlooked "extras" such as maintenance, software, peripherals, and security. Moreover, you probably won't eliminate employees. But it will provide you with new ways of doing things and will expand your capabilities in ways that may really pay off in the long run.

2.   *A computer won't make your organization run right.* If you've got problems, a computer will only make them worse. However, a computer will find better and faster ways of doing what your company does right.

3.   *A computer won't solve every problem.* Many important health and safety decisions go beyond the business computer's capability—the answers may require a good dose of subjective evaluation, something that computers cannot yet do.

4.   *A computer won't run itself.* It takes a dedicated group of individuals to make even a state-of-the-art computer run properly. Selecting that group of people can make the difference between successful and not-so-successful computing.

5.   *A computer won't always be right.* The information it puts out is only as good as the information that is put into it.

6.   *Computer security isn't automatic.* The information that you put in a computer is available to anyone who knows how to get it. Computers can be manipulated; make sure that you or your management are the only ones managing your system.

7.   *A computer needs love, too!* A well-operated computer system requires constant attention, from the temperature of its environment to its maintenance schedule. A little preventive maintenance can eliminate a lot of costly down time.

8.   *A computer won't become obsolete.* Sure, there will be new and improved computing systems, with features that are faster and more powerful than yours. But as long as your system cost-effectively provides the services your business needs, its obsolescence is only a state of mind.

PART 2
CASE HISTORIES

CHAPTER 6

SURVEYS AND INSPECTIONS OF ACADEMIC CHEMICAL STORAGE FACILITIES

PATRICIA ANN REDDEN

**Saint Peter's College, Jersey City, New Jersey and Safety
Committee of the New York Section of the
American Chemical Society**

6.1   INTRODUCTION

Many conferences, short courses, and symposia at professional meetings are devoted to solving problems involved in storing and handling chemicals. Equally active are governmental agencies such as the National Institute for Occupational Safety and Health (NIOSH), the Occupational Safety and Health Administration (OSHA), the Environmental Protection Agency (EPA), and the U.S. Consumer Product Safety Commission (CPSC) and professional groups such as the American Chemical Society (ACS) and the National Science Teachers' Association (NSTA). Books and journal articles proliferate on the question of safety in working with chemicals.

This chapter identifies the current conditions of laboratory safety in a select group of the academic community. Both industrial and academic facilities are subject to the safety regulations of governmental agencies, which conduct regular inspections to ensure the compliance that results in worker safety. However, such inspections are rare in academic institutions. In most cases, academic institutions must address the safety issue without the guidance provided by these inspections. Unsafe practices in academic laboratories result in special problems that must be recognized and corrected.

6.2   SOURCES OF DATA

The main sources of our data will be the reports of the Safety Committee of the New York Section of the American Chemical Society, surveys taken by Lab Safety Supply Company, a memorandum prepared by the Chemical Hazards Program of the U.S. Consumer Product Safety Commission, and a published survey on the results of OSHA-type inspections of academic institutions. The data from these surveys will give the reader a better understanding of the scope of these problems.

The *Safety Committee of the New York Section of the ACS* (hereafter referred to as N.Y.–ACS Safety Committee) was first formed in 1979. There are currently 13 members from academic institutions and one non-academic member. Each year a letter is sent to the colleges in the New York Section inviting them to take advantage of an on-site safety inspection. The standards used for these inspections are described in the American Chemical Society's publication, *Safety in Academic Chemistry Lab-*

oratories (1). The method and forms used in the inspections have been previously discussed in detail (2), as have the results of the first year's inspections (3).

During the spring of 1980, the Safety Committee conducted on-site inspections of eleven colleges in the New York metropolitan area. In 1981 10 of these colleges and universities were reinspected, and eight additional institutions were inspected for the first time. Two more colleges were inspected in 1982. These 21 inspected colleges are in an area that includes New York City, Westchester, part of northern New Jersey, and all of Long Island. Nine public colleges and 12 private colleges comprise the list. Four have college graduate programs, 15 grant only the baccalaureate degree, and 2 are community colleges. The number of full-time faculty in the chemistry departments generally ranges from 3 to 12, although institutions with graduate programs have considerably larger faculties. The number of students using the chemistry laboratories per week varies between 100 and 2000. The number of chemistry majors is reported as between 10 and 150.

Each institution* was asked to complete a preinspection form. Then a team that consisted of two to six inspectors, depending on the size of the college and the number of laboratories, conducted the inspection. The results were summarized and sent to the department's safety officer or chairperson with recommendations for improvement where needed. The same procedure was followed for reinspections. Careful note was taken of previously made recommendations and attempts made to implement them. The inspection committees were so thorough that, to the surprise of many of the inspected institutions, storage areas and laboratories were inspected bottle by bottle and bench by bench. Since most of the inspectors were academicians, they were particularly aware of the hazards posed by relatively inexperienced students handling chemicals and equipment. At the same time, the constraints (financial and otherwise) that handicap chemistry departments trying to make their facilities safer were clearly understood. The recommendations of the committee were made with these points in mind.

Lab Safety Supply Company personnel conducted surveys at the Pittsburgh Conference (March 1981 in Atlantic City, New Jersey) and at the

---

*Since the intention of this chapter is not to present the problems of specific colleges, but to draw conclusions about college laboratory safety that may be informative to the scientific community as a whole, the names of the colleges inspected are withheld.

National Science Teachers' Association's National Convention (April 1981 in New York City). Excerpts from these surveys were published in the *Journal of Chemical Education* (4). One hundred sixty-five questionnaires were completed at the Pittsburgh Conference. Unfortunately, there is no breakdown of the results to indicate how many of the respondents were from industry and how many were from academic institutions. The questionnaire at the NSTA convention was completed by 143 science educators, the majority of whom taught in grades 7–12. A breakdown by grade level and discipline taught is found in Table 6.1. The questionnaires themselves and the responses are given in Tables 6.2 and 6.3.

The U.S. Consumer Product Safety Commission (CPSC) conducted a limited survey through its 11 regional and district offices. Each office was requested to contact the safety officers or administrators of two school districts (one rural and one urban). Each administrator was asked to provide information on the safety equipment available in the school district, describe the methods of chemical disposal used, and list the chemicals used in the laboratory. The information was presented to the commission in a memorandum prepared by Rory Sean Fausett, Program Manager of the Chemical Hazards Program, H.S., in January 1982 (5). The memoran-

TABLE 6.1   Distribution of Survey Responses at NSTA Conference, April 1981

	Breakdown of Responses by Teaching Level	
Grade level	Number of Responses	Percent of Total Responses (%)
K–12	6	4
K–6	8	6
7–12	115	80
College	10	7
No response	4	3
	143	
	Breakdown of Responses by Subject Matter Taught	
General science	34	24
Biology	30	21
Chemistry	67	47
Other	12	8
	143	

TABLE 6.2   Survey Conducted at NSTA Conference, April 1981

Question	Yes	No
Do you have a chemical storeroom?	90%	9%
Is your storeroom normally left unlocked?	18%	79%
Does your storeroom have two or more clearly marked exits?	43%	52%
Are the aisles in your storage area free from obstruction?	66%	30%
Is the air in your storeroom dehumidified?	13%	82%
Are the shelves on which you store chemicals fastened to the wall or floor?	73%	20%
Do the bottles in which you store chemicals have labels with safety and/or first aid information?	46%	51%
Are bottles of chemicals labeled with receiving and expiration dates?	28%	69%
Are chemicals arranged alphabetically on the shelves?	73%	25%
Are chemicals arranged by class (oxidizers with oxidizers, flammables with flammables, etc.)?	47%	51%
Are chemicals stored next to a heat register, radiator, or other heat source?	4%	96%
Are stored chemicals exposed to direct sunlight?	5%	94%
Do you have a special cabinet for storing flammable liquids?	56%	41%
Do you use safety cans while storing flammables on the laboratory bench or counter?	24%	70%
Are large bottles of concentrated acids or bases stored above waist level?	13%	86%
Are bottles of concentrated acids stored separately from inorganic bases?	76%	24%
Do you use bottle carriers for transporting acid bottles?	11%	87%
Do you use neutralizing agents for spills of acids or bases?	64%	32%
Do you use absorbents for cleaning up chemical spills?	35%	61%

Courtesy of David Pipitone, Lab Safety Supply Company, Janesville, Wisconsin.

dum reports the results of a search of data base files for reports of school-related injury due to the use of chemicals in secondary schools. The files studied were the National Electronic Injury Surveillance System (NEISS), accident investigation reports, injury or potential injury reports (IPII); death certificates, and a National Safety Council publication entitled *Accident Facts*. This does not give a full picture of accidents resulting in

TABLE 6.3   Survey Conducted at Pittsburgh Conference, March 1981

Question	Yes	No
1. Are chemicals stored in a specially designated storeroom? (If not, go on to question 4)	67%	32%
2. The chemical storeroom		
a. Is locked at all times, with entry by authorized personnel only	36%	30%
b. Is identified with a sign as a chemical storeroom	42%	30%
c. Has two or more clearly marked exits	33%	30%
3. The chemical storeroom		
a. Has sufficient lighting to read labels of chemicals	65%	6%
b. Has a ventilation system that exhausts room air to the outside of the building	44%	23%
c. Has a cool, dry atmosphere (either air conditioning or dehumidifier system)	46%	22%
d. Has unobstructed aisles (i.e., no blind alleys)	48%	17%
e. Has shelving units attached to the wall or floor	52%	14%
4. All chemical containers are clearly labeled		
a. As to their contents	90%	5%
b. With receiving and disposal dates	32%	58%
5. Equipment and procedures readily available for emergencies include		
a. Approved first aid supplies	72%	20%
b. Posted emergency telephone numbers	64%	28%
c. Eyewash facilities	83%	13%
d. Drench shower facilities	81%	14%
e. Spill cleanup supplies	60%	33%
f. Fire extinguishers	92%	3%
g. Self-contained breathing apparatus	48%	44%
6. Chemicals are stored		
a. Alphabetically	63%	31%
b. By class—oxidizers with oxidizers, flammables with flammables, etc.	45%	39%
c. By random placement	23%	59%
d. In a cool, dry atmosphere	66%	25%
e. Away from direct sunlight or localized heat	87%	7%
7. Chemicals are stored in the following manner		
a. Incompatible chemicals are physically segregated	65%	24%
b. Large bottles of acids are stored on a low shelf or in acid cabinets	82%	12%
c. Oxidizing acids are segregated from organic acids and flammable and combustible materials	70%	20%

TABLE 6.3 (*Continued*)

Question	Yes	No
d. Acids are separated from inorganic bases.	70%	21%
e. Acids are separated from active metals such as sodium, potassium, magnesium, etc.	81%	10%
f. Acids are separated from chemicals that could generate toxic gases on contact: iron sulfide, sodium cyanide, etc.	79%	12%
g. Flammable liquids amounting to more than 1 pint are stored in approved safety cans or cabinets	61%	32%
h. Peroxide-forming chemicals (e.g., diethyl ether) are stored in airtight containers in a dark, cool, dry place	72%	18%
i. Peroxide-forming chemicals are labeled with receiving, opening, and disposal dates	39%	46%
j. Water-reactive chemicals such as calcium oxide are stored in a cool, dry place away from any potential water source	65%	24%
k. Oxidizers are physically segregated from flammable and combustible chemicals or materials	68%	22%
l. Oxiders are physically segregated from reducing agents such as zinc, alkaline metals, and formic acid	64%	22%

Courtesy of David Pipitone, Lab Safety Supply Company, Janesville, Wisconsin.

injuries, since only injuries severe enough to warrant hospital emergency room treatment were reported, and the NEISS coding system did not allow a national projection to be made. However, the data are of interest when combined with information about chemical handling practices in the schools.

A final source of information is a 1977 report by Schmidt (6) on the results of OSHA-type inspections of academic institutions. At that time, 19 institutions responded to a survey of chemistry department chairpersons taken throughout the United States, indicating that they had undergone an OSHA-type inspection. Nine of these inspections were carried out by state and federal OSHA, four by campus health and safety offices, one by a state Department of Labor, two by an insurance company, two by a consulting firm, and one by a visiting committee. Twelve Ph.D.-granting institutions were inspected as compared to seven undergraduate departments.

6.3   DESCRIPTION OF PHYSICAL FACILITIES

Many laboratories were built before the issue of safety was so much in the forefront as it is today. As a result, necessary safety equipment is either missing or retrofitted after construction of the laboratory. The purchase of this equipment can be a real strain on a department budget. Proper installation, maintenance, and use requires both concern for and knowledge of laboratory safety principles. Discrepancies can easily occur between what an institution or instructor believes is adequate and what is recommended by agencies knowledgeable in safety. Pertinent to physical facilities are the topics of ventilation, chemical storage facilities, personal and site safety equipment, communications and egress, and housekeeping.

6.3.1   Ventilation

Ventilation is frequently a major problem in chemical storage areas, although laboratories are normally adequately ventilated. This is particularly true of those storage areas in academic institutions that are located in a basement or a converted, often windowless cubbyhole. The ventilation in those storerooms often was not safe for the quantities of volatile chemicals that were stored. Of the 21 colleges inspected by the N.Y.–ACS Safety Committee, 11 were found to have this type of problem. In one case, the ventilation was so poor that the inspectors developed severe headaches after relatively short exposure to storeroom atmosphere. The survey taken at the Pittsburgh Conference indicates that only 44% of those responding had chemical storerooms that exhausted air out of the building, which would reduce the severity of ventilation problems.

Fume hoods in laboratories, on the whole, performed poorly in the on-site inspections. Less than half of the colleges had hoods that were all operable and working at the recommended velocity (100 cubic feet per minute), and only one college had marked the sash level at which this velocity was achieved. In fact, in several institutions the hood sashes had to be almost closed to operate correctly. Some institutions had inoperable sashes on hoods, and many had interior controls for gas, water, and so on. Adequate lighting in hoods was a problem in many cases. In one college the hood sides and sash were made of ordinary window glass rather than safety glass or wire-impregnated glass.

It is of interest to note that a list of safety problems found in the OSHA-type inspections of academic laboratories (6) listed inadequate ventilation and exhaust hood flow velocity as the fourth most frequent problem (see Table 6.4). This seems to correlate with the findings of the N.Y.–ACS Safety Committee. It is strongly suggested that each institution purchase an inexpensive vane anemometer to check the hoods for proper operation and then mark the optimum sash height for future reference.

6.3.2 Chemical Storage—Stockrooms

Although all the colleges inspected by the N.Y.–ACS Safety Committee and 90% of the educators surveyed at the NSTA conference had separate

TABLE 6.4 Safety Problems in Academic Laboratories[a]

Improper electrical wiring
  Ungrounded equipment
  Overloaded circuits
  Inappropriate high voltage shielding
Unguarded belt and pulley assemblies, saw blades, and buffer wheels
Improper storage of bulk chemicals
  Stockroom
    Design of shelves, air-handling system, fire equipment
    Ungrounded bulk solvent drums
  Laboratories—research and instructional
    Excessive volumes of solvent
    Lack of metal safety cans and metal cabinets
    Lack of explosionproof refrigerators
  Carcinogenic compounds
Inadequate ventilation and exhaust hood flow velocity
Lack of eyewash and safety shower facilities
Less major problem areas
  Waste disposal
  Storage and use of gas cylinders
  Poor housekeeping practices
  Inadequate safety signs
  Inadequate shields for individual experiments
  Blockage of escape routes between research areas

[a]Identified by OSHA-type inspections (in approximate order of the frequency of citation) (6).
Reprinted with permission from R. L. Schmidt, "Academic Experiences with O.S.H.A.," *J. Chem. Ed.*, **54**, A145 (1977).

chemical storerooms, approximately one-third of the responses at the Pittsburgh Conference indicated that chemicals were not stored in a specially designated storeroom. If separate storerooms existed, approximately half did not have two or more clearly marked exits. The inspected colleges almost invariably kept the storerooms locked, with entry by authorized personnel only, but only 79% of the responses at the NSTA conference and 50% of those at the Pittsburgh Conference with separate storerooms had a similar arrangement. Of the responses at the Pittsburgh Conference, 50% indicated that the chemical storeroom was not clearly identified.

Most of the colleges inspected by the N.Y.–ACS Safety Committee did not have adequate storage facilities for volatile or flammable chemicals. Some had "solvent rooms" with quite inadequate facilities. These were usually ordinary storage rooms filled with volatile or flammable materials. Although some colleges employed below-ground or outside bunkers, there were no provisions for maintaining a moderate temperature to minimize summer heat or winter freezing. The majority of colleges inspected, despite need, were unable to purchase needed solvent storage cabinets because of the fairly high cost. Two colleges built wooden storage cabinets conforming to OSHA standards (7,8), at a greatly reduced cost.

In the absence of suitable solvent cabinets, either fume hoods or open shelving were used for storage. Occasionally the solvents were stored in separate wooden or steel cabinets that did not conform to the safety standards. This situation was also verified by responses at the NSTA conference. Only slightly more than half of those surveyed indicated the presence of special cabinets for storing flammable liquids.

Areas in which solvents are dispensed should have facilities to ground and bond solvent cans and drums. Only three inspected colleges had such arrangements. This was also cited as a recurring problem in the report of the OSHA-type inspections.

Although large temperature fluctuations are undesirable in chemical storage areas, discussions with high school teachers indicate this to be a serious problem. Since high school laboratories seldom are used through the summer, the rooms are unventilated and may become extremely warm. One high school instructor in New Jersey reported to the author that a block of paraffin stored on an open storeroom shelf was discovered melted when class resumed in September!

A final note should be made on acid storage cabinets and explosion-proof refrigerators. Few inspected colleges had adequate storage for acids.

Most colleges stored acids on open shelves in a general storage area or in cabinets under hoods. Segregation of acids from incompatible chemicals was generally uncertain. Explosionproof refrigerators full of nonvolatile and nonexplosive chemicals indicated uncertainty about their proper usage.

6.3.3 Safety Equipment, Personal and Site

Personal and site safety equipment are frequently deficient in academic laboratories. Each laboratory should contain an adequate number of fire extinguishers of appropriate type and size, a safety shower, piped-in eyewash unit(s), spill control materials, and a minimum amount of first aid supplies, particularly sterile pads and bandages. Additionally, chemical storage areas should be equipped with automatic fire extinguishers and fire and/or smoke detectors. Self-contained breathing apparatus and materials for control of larger spills must be readily available for emergencies. Unfortunately, these supplies are often missing, inadequate, or poorly sited.

Eyewash units in particular are a problem. Several colleges and high schools depended solely on small, portable units containing approximately 500 ml of solution. Since the ACS recommendation is that the eyes be flushed for 15 minutes (1), these units are woefully inadequate and should be used only to obtain access to a piped-in eyewash In the inspected colleges, piped-in eyewash units were mounted on the cold-water line rather than supplying tepid water. Many eyewash units required the use of one hand to keep the water running, making it difficult to hold the eye open. Other units were placed so low that students would have to stoop to wash their eyes. Additionally, an inadequate number of eyewash stations often existed, so a student might have to run across an entire laboratory or half-way down a hall to use the eyewash. The lack of proper eyewash facilities is cited in the report on OSHA-type inspections and appears in the CPSC report as well.

Safety showers also have their share of problems. Every college inspected by the N.Y.–ACS Safety Committee had emergency safety showers, but these were not always adequately located or engineered. One college had a valve instead of a pull ring on the shower, which reduced the availability and the rate of delivery of the water. Another college had rings that were held over doorways by cup hooks, necessitating an upward movement before a downward pull. (Interestingly, this arrangement was approved by an industrial safety consultant, the local fire department, and the insurance company's inspectors. Only the academic members of the N.Y.–ACS Safety Committee pointed out that this would be a real hazard for students inexperienced in an emergency.) Other colleges had showers that were jointly used by two or more instructional laboratories, rather than one shower per laboratory. Even when showers were adequate, there was only one college that indicated when the showers had last been checked; most were not checked at all. Only six of the school districts responding to the CPSC survey said that they had safety showers, and this lack was also cited for the OSHA-type inspections.

The American Chemical Society recommends that each laboratory bench be equipped with one fire extinguisher of appropriate size and type (1). This might seem somewhat excessive to many chemists, and in actuality the New York–ACS Safety Committee evaluated the number of extinguishers on the basis of room size, geometry, and extinguisher type and size. Despite this reduced requirement, few inspected laboratories had adequate coverage. These deficiencies were particularly noticeable in storage areas, even though it is these areas that present the greatest potential hazard. Many extinguishers in laboratory areas were poorly marked and not mounted accessibly. Others lacked seals or were partially discharged, making them undependable in an emergency. Many sealed extinguishers did not carry tags with current dates of inspection.

Laboratory and storage areas often have no automatic alarm and detection system for smoke and/or heat. If such a system is present, it frequently rings only within the department or the building, not at a central security desk or local fire department. Isolated chemical storage areas in particular should have not only a detection–alarm system, but adequate automatic fire extinguishers.

Figure 6.1 illustrates the type of poor laboratory planning that was found in one institution inspected by the N.Y.–ACS Safety Committee. The placement of the combination safety shower–eyewash unit at the back of the laboratory between two hoods limits access by those students working at distant tables. The constricted space also makes it difficult for an instructor to administer aid to an injured student using this equipment. The college would be better off with two eyewash units mounted on the sinks of alternate laboratory tables, at the end opposite the hoods. The partially discharged fire extinguishers shown in Figure 6.1 were poorly sited for maximum accessibility.

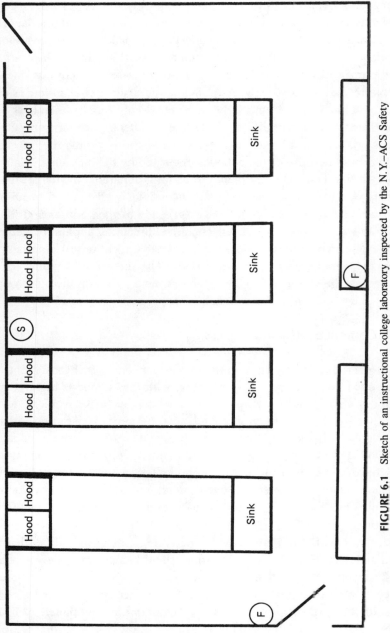

FIGURE 6.1 Sketch of an instructional college laboratory inspected by the N.Y.–ACS Safety Committee [S = combination safety shower/eyewash unit; F = fire extinguisher (wall mounted)].

Every laboratory should have a policy requiring the use of safety goggles or glasses. Most states mandate such protection in academic laboratories. However, when touring academic laboratories, one will invariably discover some students and faculty members working in the laboratory without proper eyewear. The use of prescription glasses without eye shields is often allowed, and the ACS recommendation against contact lenses (1) is usually acknowledged but rarely enforced. Many faculty members report resistance to the use of goggles by their students, primarily because of poor fit or scratched lenses that result in a visibility problem.

Only about one-third of the colleges inspected by the N.Y.–ACS Safety Committee had adequate emergency breathing apparatus available. This apparatus was not always stored in an accessible, well-marked location. The apparatus found most often were emergency oxygen bottles and filter masks. Only three colleges had self-contained breathing apparatus located outside the chemical storerooms. The predictable location for emergency breathing apparatus was in the storeroom. The danger of keeping such equipment in the storeroom occurs when an emergency in the storeroom prevents entry without a proper breathing apparatus.

6.3.4   Communication and Egress

In case of an emergency in a storage area (especially in a basement or bunker) or in a laboratory, communication with the outside is imperative. In many cases the college inspections revealed that there were no telephones nearby. This problem can be solved by simply installing an in-house telephone line or an intercom, in the event that a school is reluctant to have telephones that may be used by unauthorized persons. This phone should have a sticker or a nearby prominent sign listing the following emergency phone numbers: fire department, hospital emergency room, on-premises medical staff, poison control center, and emergency chemical information.

Emergency exits from laboratories and storage areas should be clearly marked and accessible. Ideally each room should have two widely spaced exits. Inspection revealed that the second exit was frequently either blocked or locked for security reasons. Where a second exit does not exist, consider a clearly marked emergency escape ladder or kick-out panel. At least one New Jersey high school painted bright yellow paths to the nearest egress directly on the floor, so that students could exit in the correct direction from a smoke-filled room.

6.3.5 Housekeeping

Conditions in storage areas were not very good. Shelves were too high and aisles too narrow for safely removing and carrying chemicals. Labels on bottles were missing or deteriorating. Large bottles stored on high shelves often protruded over the edges of the shelves. Metal shelves were frequently rusting. Bottles, solvent cans, and drums were piled two and three deep. These large quantities were excessive for the size of the room, and the floor area was covered, access to exits was blocked, and doors were prevented from being opened or closed. Finally, spill control was a problem with poorly labeled and hard-to-find neutralizing and containment materials.

These problems also appear in the report on the OSHA-type inspections of academic institutions.

6.4 CHEMICAL STORAGE AND DISPOSAL

6.4.1 Inventory Control

As student enrollments in laboratory courses drop, faculty turnovers occur, and laboratory syllabi change, the need for chemicals varies. However, ordering practices do not always reflect these changes. Thus large amounts of chemicals accumulate in quantities that cannot be used even over several years. In many cases these chemicals are hazardous or possess a limited shelf life. One college inspected by the N.Y.–ACS Safety Committee had a 65-lb container of ammonium nitrate in a solvent storage room. Another had over 50 one-gallon bottles of glacial acetic acid on its shelves. Large quantities of flammable solvents were found in several colleges. Diethyl ether, known to form explosive peroxides, was often undated or out of date. Another college had been "blessed" with a donation of fuming acids, alkali metals, and lachrymators some years ago and is now desperately trying to store and dispose of them properly. Economics has been a reason for ordering large quantities of chemicals in that a better unit price can be obtained. Four of the school districts in the CPSC survey indicated that chemicals were ordered in large volume to cut costs. Sixteen of the school districts store chemicals from year to year rather than order only what will be used for the current school year.

Proper inventory control can often minimize the excesses. A system of dating materials on arrival encourages the use of older chemicals first. Only 28% of those completing surveys at the NSTA conference and 32% of those at the Pittsburgh Conference indicated that chemical containers were clearly marked with receiving and disposal dates. This is of particular importance for ethers and peroxide-forming chemicals, which should also be marked with the date they were opened (9).

6.4.2   Compatibility Storage

Compatibility storage costs nothing but time. The N.Y.–ACS Safety Committee found that only a few of the 21 inspected colleges attempted to separate stored chemicals according to their reactivities. In only three cases was the segregation respectable. Most storage rooms segregated only strong acids from concentrated bases or kept organic compounds separate from inorganics. Flammable or volatile organic solvents, unless in drums or cans, were usually arranged alphabetically among other organic compounds. Bromine and sodium metal were often stored on common shelves. Bromine and nitric acid, both strong oxidizing agents, were frequently in close proximity to organic chemicals. Concentrated hydrochloric acid and ammonium hydroxide were found side by side. The survey responses in Tables 6.2 and 6.3 indicate the universality of this problem.

Figure 6.2 illustrates the magnitude of the problem when many of these individual hazards are combined in a chemical storeroom. This is a sketch of a storage area found in one of the colleges inspected by the N.Y.–ACS Safety Committee. Trouble spots are indicated.

There is no good reason why storage of chemicals according to compatibility cannot be achieved. The most common reason given for not storing chemicals according to compatibility deals with the lack of compatibility information. However, there is a list of incompatible chemical combinations in the appendix of the ACS manual (1). Pipitone (10) has created a checklist that can be used as a starting point to achieve good storage practice. In addition, the manufacturers' labels on chemical containers give a good deal of information about reactivities and special storage needs (11). Another good source for information on both correct storage and disposal methods is the chemical catalog reference manual of Flinn Scientific, Inc. (12). Many articles and books have appeared describing inventory methods and labeling techniques to convey information about safety, storage, and disposal (4,13–15).

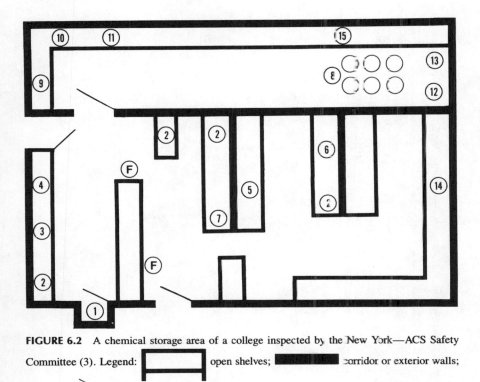

FIGURE 6.2 A chemical storage area of a college inspected by the New York—ACS Safety Committee (3). Legend: ▭ open shelves; ▮▮▮ corridor or exterior walls; ▮▮▮ doors; Ⓕ fire extinguisher. Key: 1, closet containing potassium metal piled haphazardly among cardboard shipping cartons filled with assorted chemicals; 2, unlabeled bottles or bottles with illegible labels; 3, chemicals spilled on shelves; 4, strong oxidants ($KBrO_3$, $KClO_3$) on shelves with oxidizable chemicals; 5, concentrated acids and bases stored together; 6, chromic acid on shelves with organic compounds; 7, bottles of 30% hydrogen peroxide next to concentrated acetic acid; 8, 5-gallon cans of organic solvents piled two and three deep; 9, bromine and benzene in alphabetical order; cans corroded; 10, anhydrous ether, undated cans; 11, collodion (extremely flammable) on open shelves; 12, diethyl ether in randomly piled boxes, approximately 5 feet high; expiration date 18 months prior to inspection; 13, box of chemicals labeled "potent carcinogens"; 14, large bottles on edges of top shelves (approximately 10 feet high); 15, several bottles labeled only as "organic waste."

6.4.3 Toxic Substances and Carcinogens

In its survey, the CPSC requested a list of chemicals used in the 22 secondary school districts. The total list included 312 chemicals. No attempt was made to differentiate between those actually in use and those kept in stock but not used. The hazardous properties (if any) of these

TABLE 6.5 Compounds Identified by the U.S.CPSC Report as Carcinogenic, Teratogenic, Mutagenic, or Causing Neoplasm Formation (5)[a]

| | Identification by Sax (16) | | | | Identification by STIC | | |
	Carcinogen	Teratogen	Causes Neoplasms	Mutagen	DHHS	EPA–CAG	OSHA
Acetamide	exp (+)						X
Aniline hydrochloride	exp (±)						X
Benzene	Recognized leukemogen				X	X	X
Benzidine	recog				X	X	X
Cadmium chloride	exp (+)	exp	exp		X	X	
Cadmium nitrate					X	X	
Carbon tetrachloride	exp (+)	exp	exp (S)				
Chloroform	exp (S)		recog			X	
Chromic acetate	recog						
Chromic acid	exp (+)						
Colchicine	exp	exp	exp	exp			
Dichloromethane	exp						
1,2-Dichlorobenzene	exp (±)					X	
2,4-Dichlorophenol	exp						
Diphenylamine	exp	exp					
Ethylene dichloride	exp	exp		exp			
Ferric oxide	exp (±)						
Formaldehyde	exp					X	
Isoamyl alcohol	exp						
Isobutyl alcohol	exp						
Kerosene	susp						

146

Lauric acid						exp
Lead acetate	exp (+)	exp				
Lead chloride		exp				
Lead nitrate		exp				
Lithium chloride		exp				
Methyl ethyl ketone		exp				
Nickel(ous) ammonium sulfate			X	X		
Nickel(ous) chloride			X	X		
Nickel(ous) sulfate			X	X	exp	
Phenol	exp					
1-Propyl alcohol	exp					
Pyrogallic acid	exp					
Salicylamide		exp			exp	
Sodium chromate	exp (±)			poss		
Sodium dichromate	exp (S)					
Tannic acid	exp (+)					
Thioacetamide	exp (+)					X

[a]*Symbols used*: exp, experimental; poss, possible; recog, recognized; susp, suspected; (+), on review by International Agency for Research on Cancer (IARC), classified as a carcinogen; (S), on review, IARC classified it as a suspected carcinogen; (±), on review, IARC states there is insufficient data for classification; X, identified by the group named as a carcinogen (potential or recognized).

chemicals were assessed using *Dangerous Properties of Industrial Materials,* 5th ed., by N. Irving Sax (16); System for Tracking the Inventory of Chemicals (STIC); the *NIOSH/OSHA Pocket Guide to Chemical Hazards* (15); and a commercial chart, "Toxic and Hazardous Chemicals in Industry," prepared by Science Related Materials, Inc. (17). The STIC is a new CPSC database using listings of carcinogens by OSHA, EPA's Carcinogen Assessment Group (CAG), and the Department of Health and Human Services' (DHHS) first *Annual Report on Carcinogens*.

Table 6.5 lists those chemicals in the schools' inventories classified by Sax or the STIC database as potential or recognized carcinogens. There is not universal agreement on these classifications. Sax identifies 28 potential carcinogens and OSHA only 4. Thirteen of those listed by Sax have been reviewed by the International Agency for Research on Cancer (IARC). Of these, seven were classified as carcinogens and two as suspected carcinogens. For the remaining four there was insufficient data to make a proper classification. Eleven of the chemicals were identified as potential teratogens, whereas two are possibly mutagenic. It is interesting that benzene is on the list. Many older laboratory manuals ascribed benzene as a standard solvent. Newer editions of the manuals reflect a change to other solvents. However, high schools often use the same editions for several years after publication since the books are purchased by the school.

The presence of potential carcinogenic chemicals in the surveyed secondary school districts is evident from the findings of the OSHA-type inspections (see Table 6.4) and from the inspections by the N.Y.–ACS Safety Committee (cf. item 13 in Figure 6.2). Many institutions are aware of this problem and are taking steps to either eliminate or safely identify and store carcinogens. Others indeed seem oblivious to the issue.

Carcinogenicity is not, of course, the only hazard identified with chemicals. Compounds such as linseed and cottonseed oils (allergens), hydrosine sulfate (a poisonous alkaloid), and lithium hydroxide (very caustic and toxic) are available to students, often with insufficient warning. The "Toxic and Hazardous Chemicals" safety chart used by CPSC rates chemicals according to toxicity or health hazard, flammability, and chemical reactivity on a scale of 0–4. A health hazard rating of three means that short-term exposure may result in major temporary or permanent injury and may threaten life. A rating of 4 means that short-term exposure may result in major permanent injury or death. In the CPSC survey, 31 chemicals had a health hazard rating of 3 and two had a rating of 4. Recognition of the hazard and careful usage minimize the risks for many of these compounds, whereas some should definitely be eliminated from an academic institution's inventory.

A *Manual of Safety and Health Hazards in the School Science Laboratory* (18) has recently been published by NIOSH. This manual evaluates the hazards connected with chemicals, equipment, and procedures in experiments typical of high school courses in chemistry, biology, earth science, and physics. Each chemical is assigned a health rating as those on the previously mentioned safety chart. In addition, an appendix identifies chemicals with greater hazardous nature than potential usefulness, chemicals that should be removed from the schools if alternatives can be used, or chemicals that should be retained only in minimal amounts. Detailed information is given on the toxicity, dose levels, and routes of exposure of carcinogens (positive, potential, or suspected) and of other toxic compounds.

6.4.4 Spill Control

The topic of spill control must be addressed on both large and small scale, for spills of both solids and liquids (4).

Solid materials may be cleaned up fairly easy with a dustpan and brush, but their treatment after cleanup should be carefully considered. Dumping a strong oxidant such as potassium perchlorate in a wastebasket full of paper is certainly not a recommended procedure as fires can occur! Compatibility should be considered when several chemicals are being placed into the same waste container. The shelves in several of the storage areas inspected by the N.Y.-ACS Safety Committee contained spilled chemicals that had not been cleaned. Many storage areas did not have containers for collection of solid waste.

Liquid spills are a more treacherous problem than solid spills because of seepage into cracks, flow properties that can carry the material under doors and cabinets, and volatility. The liquid spilled must often be neutralized as well as removed. Traditional methods of controlling liquid spills include the use of sodium bicarbonate to neutralize acid spills and diatomaceous earth, sand, or vermiculite to absorb organic materials. Currently available commercial products include pillows of various sizes, filled with an absorbent that can absorb up to 98% of the pillow's capacity in less than 30 seconds. The whole pillow, with the absorbed contents, can then be disposed of correctly and safely. Deep trays under bottles and sills in solvent storage rooms to contain spills are also recommended.

The survey at the NSTA conference showed that 64% of the respondents used neutralizing agents for acid or base spills, but only 35% used other absorbents. At the Pittsburgh Conference, 60% of those surveyed used spill cleanup supplies (undifferentiated by type). The N.Y.–ACS inspections also identified a lack of adequate materials in many of the inspected colleges.

A final point should be raised on this topic. Chemical spills may cause moderate to severe burns to the person coming into contact with the material. Neutralizing materials such as sodium bicarbonate paste or solution (for acid burns) and sodium thiosulfate solution (for bromine spills and burns) should be placed adjacent to the area where these chemicals are to be used and should be prominently labeled for emergency first aid. This is particularly important when a minor spill down the side of a reagent bottle may be temporarily undetected until an accident occurs.

6.4.5 Waste Disposal

Disposal of waste chemicals is a major problem confronting chemical laboratories, particularly academic laboratories. Each college inspected by the N.Y.–ACS Safety Committee was in the process of developing its own method for the safe disposal of chemical wastes. This leads, however, to the practice of keeping unlabeled bottles and waste solvents in storage areas since safe, legal disposal techniques are currently too expensive for many high schools and colleges. Some colleges have found that their local fire departments or bomb squads are willing to help remove their waste. Another college recovers some of its solvents by fractional distillation. Unfortunately, two of the colleges surveyed have been dumping the waste down the drain and depending on acid traps to neutralize the liquids. Acid traps have no effect on the organic solvents that are disposed of in this manner.

The CPSC survey of chemicals in secondary school laboratories reveals that 20 of these chemicals are classified as hazardous wastes by EPA's Resource Conservation and Recovery Act (RCRA). A listing of these chemicals can be found in Table 6.6. Table 6.4 indicates that proper disposal is a problem in the institutions which underwent OSHA-type inspections.

TABLE 6.6 Compounds Identified on U.S.CPSC Survey of
Chemicals in Secondary Schools That Are on the RCRA List

Acetone	Lactic acid
1-Butanol	Methyl methacrylate
Benzene	Naphthalene
Benzidine reagent	Nickel(ous) ammonium sulfate
Cyclohexane	Nickel(ous) chloride
Carbon tetrachloride	Nickel(ous) sulfate
Chloroform	Phthalic anhydride
Dichloromethane	Thioacetamide
Formic acid	Toluene
Isobutyl alcohol	Xylene

6.5 CONCLUSIONS AND RECOMMENDATIONS

This chapter has emphasized problem areas from the data in surveys and inspections in laboratories and storage areas. There are many such problems. However, it would be remiss of the author not to point out the positive signs as well. The inspections of the N.Y.–ACS Safety Committee and the surveys referred to in this chapter indicate that many academic institutions and chemical companies are making a serious effort to improve their safety posture and are indeed succeeding. The participation of academic institutions in voluntary inspection programs is an obvious example of this trend, as is the attendance of faculty members and chemists at safety workshops and symposia. The continuation and expansion of these programs is essential.

Inspection of academic laboratories and storage areas, by an external professional group (as in the case of the N.Y.–ACS Safety Committee) with a knowledge of institutional needs and problems, can point to the ways and means of improvement. Most of the colleges inspected used the committee's reports as support for financial requests by the department for needed safety equipment. Problems that had been overlooked even by hard-working safety officers and committees were revealed by the inspectors' visit. As a result of the first year's inspection reports, all but one college showed substantial improvement after reinspection 1 year later. The greatest improvements were made in the area of safety equipment and housekeeping. Piped-in eyewash units, proper safety showers, adequate fire extinguishers, and safety signs were relatively easy additions to the laboratory. Housekeeping problems were also easily remedied. In contrast, neither chemical storage nor structural problems were addressed in a majority of cases. The major reason is the lead time needed to complete the work. The acquisition of funds was not possible in such a short time period.

ACKNOWLEDGMENTS

The author thanks the New York Section of the American Chemical Society and the society's Program Development Fund for their help, both financial and moral, in setting up the Safety Committee and supporting its work. Gratitude is expressed to the safety officers and faculty members of the inspected colleges for agreeing to the publication of our results. Finally, gratitude is due to the members of the Safety Committee who have donated many hours of time, both on inspections and in preparing the following reports to the colleges.

The author also wishes to thank Mr. David Pipitone of Lab Safety Supply Company and the staff members of the U.S. Consumer Product Safety Commission's Chemical Hazards Program for providing the full results of their surveys for this summary.

REFERENCES

1. *Safety in Academic Chemistry Laboratories,* 3rd ed., American Chemical Society Committee on Chemical Safety, Washington, DC, 1979.

2. Safety Committee of the New York Section of the American Chemical Society, "Guidelines for a Complete Safety Audit in the Chemistry Laboratory," *J. Chem. Ed.,* **58,** A361 (1981).

3. Safety Committee of the New York Section of the American Chemical Society, "Results of Safety Inspections of College Laboratory and Chemical Storage Facilities," *J. Chem. Ed.,* **59,** A9 (1982).

4. D. A. Pipitone and D. D. Hedberg, "Safe Chemical Storage: A Pound of Prevention Is Worth a Ton of Trouble," *J. Chem. Ed.,* **59,** A159 (1982).

5. R. S. Fausett, "Status Report: School Laboratory Chemicals," Memorandum to the United States Consumer Product Safety Commission, Washington, DC, January 5, 1982.

6. R. L. Schmidt, "Academic Experiences with O.S.H.A.," *J. Chem. Ed.,* **54,** A145 (1977).

7. *Federal Register,* **39,** No. 125, June 27, 1974.

8. N. V. Steere, "Fire-Protected Storage for Records and Chemicals," *Safety in the Chem-*

istry Laboratory, Vol. 1, Division of Chemical Education of the American Chemical Society, Easton, 1967, pp. 44–47.

9. H. L. Jackson, W. B. McCormack, C. S. Rondestvedt, K. C. Smelz, and I. E. Viele, "Control of Peroxidizable Compounds," *J. Chem. Ed.,* **47,** A175 (1970).

10. D. Pipitone, "Safe Storage of Chemicals," *The Science Teacher,* **48** (2) (1981).

11. A. J. Shurpik and H. J. Beim, "A Chemist's View of Labeling Hazardous Materials as Required by the U.S. Department of Transportation," *J. Chem Ed.,* **59,** A45 (1982).

12. Flinn Scientific Inc., P.O. Box 231, 910 W. Wilson St., Batavia, Ill. 60510; Chemical Catalog Reference Manual, issued annually.

13. M. E. Green and A. Turk, *Safety in Working with Chemicals,* MacMillan, New York, 1978.

14. N. V. Steere, *Handbook of Laboratory Safety,* 2nd ed., Chemical Rubber Publishing Company, Cleveland, 1971.

15. F. W. Mackinson, R. Stricoff and L. J. Partridge, Jr. (Eds.), *NIOSH/OSHA Pocket Guide to Chemical Hazards,* U.S. Government Printing Office, Washington, DC, 1978.

16. N. I. Sax, *Dangerous Properties of Industrial Materials,* 5th ed., Van Nostrand Reinhold, New York, 1979.

17. Laboratory Safety Supply Co., P.O. Box 1368, Janesville, WI 53545; catalog number 20-2011 (pocket size, catalog number 20-2012).

18. Division of Training and Manpower Development (NIOSH), *Manual of Safety and Health Hazards in the School Science Laboratory,* U.S. Department of Health and Human Services, Cincinnati, 1980.

CHAPTER 7

THE UNIVERSITY OF AKRON CHEMICAL STORAGE FACILITIES

F. L. CHLAD
University of Akron, Akron, Ohio

7.1   SAFETY AND FACILITIES

One of the greatest problems facing chemistry departments is that of providing adequate and safe chemical storage facilities. All too often this is an afterthought, rather than being an integral part of the basic planning and design of a new building.

The central storage facility for chemicals must be carefully thought out and well planned, and there can be no doubt that it will be expensive if done properly. Among the many critical areas that need to be considered are the following:

1. An adequate space allocation, including future growth potential
2. An efficient ventilation system
3. Heat and smoke detectors
4. An automatic fire suppression system
5. Explosionproof lighting (and any motors)
6. Static-free light switches and electrical outlets
7. Proper temperature and humidity controls
8. Pressure-releasing plastic blow-out panels
9. Fire alarm and annunciator system

One must not view chemical storage facilities as some type of appendage, separate and distinct from the chemistry building itself. Rather, they should be an extension of the safe design concepts that must be planned into the total building. The ultimate goal is to plan a totally safe building, with equal consideration given to the chemical storage facilities during the original planning and design.

For purposes of illustration, this author will utilize the facilities at the University of Akron (Akron, Ohio) as an example of the total safety design concept.

Knight Chemical Laboratory, which houses the Department of Chemistry at the University of Akron, has received worldwide acclaim as being one of the safest academic chemistry buildings ever constructed. Now in its fifth year of operation, it still remains a prototype for illustrating what can be accomplished when safety is the dominant theme in designing chemical facilities. Well over 100 inquiries have been received from all

FIGURE 7.1   Knight Chemical Laboratory at the University of Akron, Akron, Ohio. External fume hood exhaust chases are shown.

over the world, and thus far 21 institutions have sent representatives to view this structure (Figure 7.1).

7.2   LABORATORY FACILITIES

One of the more important of the many innovative safety features incorporated into the laboratories was the installation of 166 induced-air fume hoods, which makes it possible for *all* experiments to be conducted in hoods. To conserve energy, the hoods take only 25% of the air from inside the rooms. The remaining 75% is tempered outside air drawn into the building at each floor level. The facility is odor-free because chemicals fumes are drawn out of the building at approximately three times the rate of exhaust obtained with a standard-type ventilation system.

7.2.1   Undergraduate Laboratories

Generally, the undergraduate organic laboratory poses the largest potential danger as far as accidents, fire, and noxious fumes are concerned. This is due to both the nature of the organic chemicals utilized and the use of large quantities of solvents. An in-depth study of accidents ascertained that the largest number of accidents occurred in the undergraduate organic laboratory. These incidents ran the full spectrum from minor fires caused by exposing solvent vapors to bunsen burners, to chemical burns caused by carrying the materials from a laboratory bench to the hood, and bumping into someone. Our new design and procedures have improved the situation tremendously.

The organic laboratory is divided into two areas served by a common instrument room. Each of the two laboratories contains 16 8-foot auxiliary air hoods at which the students do *all* of their work. There are two student stations per hood, giving a total of 64 working stations at any given time. Because all work is done in the hoods, rather than at benches and hoods, the previous congestion and safety problems associated with this laboratory have been greatly minimized (Figure 7.2).

Because of the large number of hoods in the two undergraduate organic laboratories, a tremendous amount of outside air is brought in to maintain

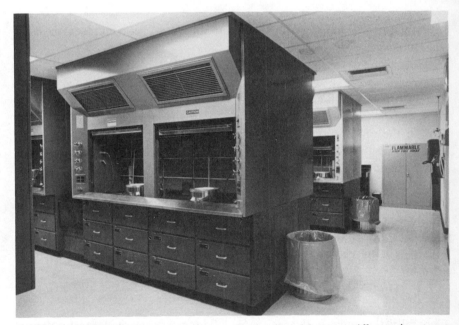

FIGURE 7.2 A standard undergraduate organic chemistry laboratory. All experiments are conducted in fume hoods.

the 75:25 ratio that is exhausted. As an example, rooms of this size without the hoods would normally require 4–6 air changes per hour, whereas under the system designed for us, the laboratories receive 109 air changes per hour, or one every 33 seconds.

Several other factors are noteworthy. The organic laboratory is now flameless since bunsen burners have been replaced with heating mantles, steam baths, and hot plates. We have converted to 19/22 standard taper semimicro glassware kits for all students. This has reduced the quantities of materials used in experiments, resulting in an economic and safer operation.

Compare, if you will, the obvious safety features described above with the situation that prevailed in the "classic" organic laboratory. The under-graduate laboratory, we all too vividly recall, served 24–36 students and forced the students to share four to six ill-working hoods that were nor-mally located against one wall. For each person working in the hood there were three or four waiting in line, all carrying their materials and chemi-cals from the bench to the hood and back again. Many didn't bother waiting for the hood to be free and did their work on open benches. Solvent fires caused by vapors reaching a lit bunsen burner were all too common. Designing safety into the planning of these laboratories has allowed us to make tremendous progress.

The concept that all chemical work must be performed in a controlled environment is also carried out at the freshman level There are eight introductory chemistry laboratories arranged in four blocks of two, with each pair of two separated by sliding pocket doors. A laboratory has 24 student stations, each one with a bench-top "T"-type hood. The hood has an adjustable baffle to allow for the removal of vapors heavier or lighter than air. Each laboratory has, in addition, an induced-air fume hood from which chemicals are dispensed during each laboratory period. The 24 "T" hoods are exhausted as a group through a common duct, whereas the storage hood is exhausted separately (Figure 7.3).

7.2.2 Graduate Laboratories

Much consideration was given to the design of the graduate research laboratories. Each laboratory is 24 × 30 feet and is designed for occu-pancy by four students. It was felt that installation of a series of two-student laboratories would result in a needless duplication of equipment,

FIGURE 7.3 Introductory chemistry laboratory illustrating the individual "T"-type hoods at each station. In case of emergency, each laboratory can be quickly evacuated.

whereas placement of more than four people in a single room would lead to overcrowding and a lack of identifiable space. Our experience with hood usage in our old building revealed that most experiments were carried out on bench tops because the hoods were used primarily for chemical storage. This practice had the further disadvantage that it required the hoods to be left on continuously. Our solution to this in the new facility was to place five hoods in each four-student laboratory. Four of the hoods are used to conduct experiments, whereas the fifth hood is used solely for storage. The four "working" hoods are turned on only when in actual use, and the storage hood is left on continuously. The storage hood does not have any utilities piped to it and is fitted with shelves. The net result is that the working hoods are not cluttered with chemicals and are turned on only when a reaction is running. Thus conditions are safer and energy is conserved as well.

The location of the graduate research student desks within the laboratory is vitally important. The ideal situation is to have desks located against interior walls to allow quick egress to the hallway. The hoods are located against exterior walls so that ducting into chases on the outside of the building may occur. These measures allow a student to exit from the laboratory quickly in an emergency situation and also eliminates the necessity of having other students or visitors walk through a potentially dangerous area to reach a graduate research student for conferences. The location of the hoods against the outer wall also eliminates the turbulence that would be caused by opening and closing doors and when persons walk continually past the hoods.

Each four-student graduate research laboratory has a "safety island" consisting of an approved solvent storage cabinet, fire extinguisher, water sprayer, bucket of sand, eyewash station, and spill kit (Figure 7.4). Another key feature in the research laboratories is that the utilities in each

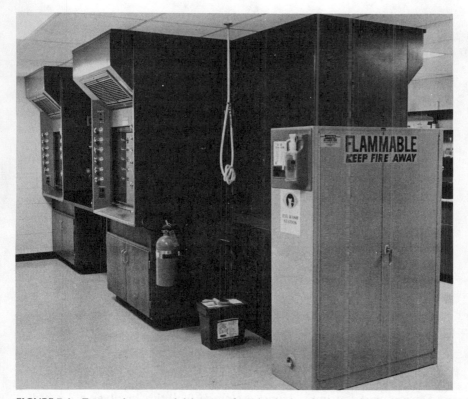

FIGURE 7.4   Four-student research laboratory featuring back-to-back "work" hoods, a storage hood, and a "safety" island.

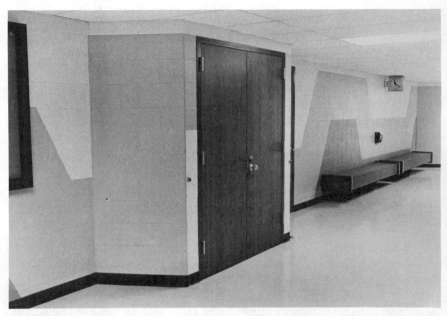

FIGURE 7.5 An access closet located outside each laboratory contains utility shut-off valves, for emergencies.

individual laboratory are able to be shut off from a control panel located in the corridor directly outside each room, so that in case of fire or other accident any laboratory can be quickly isolated from the others (Figure 7.5).

7.3 FUNCTIONAL CHEMICAL STORAGE

Because of the diverse nature of chemicals and their use, the safe storage of chemicals has become a matter of paramount importance. Ideal hazardous chemical storage would entail the complete isolation of each hazard category and even isolation of some materials within a class. However, from a practical standpoint, such isolation is not often economically feasible, and it is thus necessary to group items so that whatever space is available is used in the safest possible manner.

Designing safety into chemical storage facilities involves a great deal more than sketching out a proposed floor plan and specifying certain safety "hardware." It requires a careful examination of the total concept of chemical safety from a logistical standpoint, with a complete overview of what happens to a chemical from the time it is received at the loading dock, until ultimate consumption, or disposal of the chemical as a waste.

7.3.1 Chemical Storage and Undergraduate Laboratories

In all too many instances the following examples are common practice. A chemical ordered for a particular undergraduate teaching laboratory is received and is placed in that teaching laboratory until it is used. Each undergraduate laboratory thus becomes a "ministockroom" with little or no control procedures. Particularly in the case of the undergraduate organic laboratories, the toxic or noxious chemicals are stored in fume hoods. The fume hoods become "storage" areas rather than "working" areas. Normally, there is not a sufficient number of hoods in which to work.

Another situation occurs frequently when experiments or laboratory manuals are changed in the curriculum. Many chemicals that are no longer needed as a result of the change are still stored in the laboratory with the thought that undoubtedly usage will occur "sometime" in the future. This causes untold problems with storage space, the reliability of the chemical, and ultimately, a costly disposal situation.

The Department of Chemistry at the University of Akron has adopted a policy that has worked exceedingly well. Absolutely no chemicals are stored in undergraduate laboratories. Instead, the department has several dispensing stockrooms that service the various teaching laboratories by means of a "cart" system. The system works in the following manner. The dispensing stockroom storekeeper requisitions those chemicals that will be needed for the following week's laboratories from the main chemical storage area. The storekeeper then prepares the needed unknowns, solutions, and materials that will be required. These items are placed on laboratory carts, each identified by the course and laboratory room number. The graduate teaching assistants obtain these chemicals at the dispensing stockroom 5 minutes prior to the start of the actual laboratory period.

These materials are wheeled into the individual laboratories and are then issued to the students. At the end of the laboratory period any unused materials or waste is loaded onto the cart and returned to the dispensing stockroom. Several objectives are accomplished by utilizing this system: (a) there is a high degree of control exercised over the use of the chemicals; (b) laboratory maintenance is excellent since chemicals are absent in the laboratories when not in actual use; and (c) the administration is able to analyze accurately the cost of operation for each course. The end result is an uncluttered laboratory with no storage problems and, most importantly, a safe operation.

Our philosophy is based on the premise that if one has an excellent chemical storage facility, with all the many safety features mentioned at the beginning of the chapter, one should store the chemicals in that area for as long a period of time as possible prior to their actual use.

7.3.2 Chemical Storage and Graduate Laboratories

The problem of chemical storage in research laboratories is also a major concern. In many respects this is even more complex because one needs to deal with a host of faculty conceptions, each presenting a varying degree of safety consciousness. In addition, some faculty with funded research often feel that they are "independent" of any department system of safety or inventory control. The department must, however, have a firm policy of disallowing private "caches" of chemicals in the research areas and should limit use to only those chemicals that are currently being utilized. Everyone should follow one set of guidelines, with equal enforcement.

All too frequently, a research group will sign out a 1-lb bottle of a substance, use 50 g in a reaction that does not work, and then store the remainder on a shelf in their laboratory for years. One can easily envision the large quantity of chemicals that can be accumulated over a short period of time. Laboratories should be "work" areas, and not "storage" areas.

Our solution to this situation was the establishment of a chemical "morgue" that operates in the following manner. Once a chemical is determined to be in excess, it is brought to the main chemical storage facility where it is properly catalogued and placed in the chemical morgue. This morgue area is distinctly separate from the regular chemical stock and is kept on a different inventory system. Any faculty member or graduate research student who has a need may sign for this chemical at no charge. In this manner we are able to recycle all excess chemicals and keep supply costs to research grants and the department budget at a minimum. Storage conditions are improved from both a safety and space standpoint, and ultimately waste chemical disposal is kept to the lowest possible level.

The use of a morgue system does not have to be limited to only academic institutions. The author has had occasion to visit many industrial research laboratories. A good many of these were cluttered and under-hooded and had unsafe quantities of chemicals stored in their work areas. These conditions represent very high potential for disaster.

What is being discussed here is the classic trade-off between convenience and safety: the convenience of having that "ministockroom" in the research laboratory with an unsafe condition, or the willingness to make a few extra trips per week to the central storage facility, with a safe situation. The amounts of toxic, flammable, unstable, or highly reactive materials that should be allowed to be stored in individual laboratories must be governed. Unrestricted quantities of these materials can bring about untold safety problems and hazards. Ground rules should be established and then conscientiously enforced. An excellent reference for determining the maximum quantities of flammable and combustible liquids in laboratory units outside of approved flammable liquid storage rooms is NFPA booklet No. 45-1981, *Fire Protection for Laboratories Using Chemicals* (prepared by the National Fire Protection Association).

7.4   CHEMICAL STORES AND WASTE DISPOSAL

With the advent of the Resource Conservation and Recovery Act, tighter control of chemical inventory and the subsequent utilization of this morgue system has become a matter of extreme importance. When our chemistry department moved from the old building to our new facility, the magnitude of the problem of poor inventory control of chemicals became very evident. Many of the laboratories, both teaching and research, had a 15–20-year accumulation of chemicals. A determination had to be made as to whether these chemicals could still be utilized or whether disposal was needed. Approximately 25% of the chemicals that had been stored were discarded. In this group, chemical containers had no labels or labels that could no longer be read, container caps had corroded completely, and chemicals had become contaminated.

The identification process and ultimate disposal was a nightmare. The cost in work-hours and dollars, in addition to the highly hazardous conditions that were uncovered, were definitely instrumental in the formulation of our new policies and procedures.

Another situation resulting from the Resource Conservation and Recovery Act caused an additional problem situation. Our department was deluged with the offer of "free" chemicals. A host of companies and high schools contacted our department to offer "gifts" of chemicals claimed to be in excess to their operation. In reality, they were trying to unload old, and in many cases contaminated, chemicals that they realized needed to be discarded. By offering these chemicals as "gifts," these people not only hoped to avoid high disposal costs, but were also seeking a tax deduction for a charitable contribution. We set a policy for accepting donations of chemicals that includes a stipulation that we be allowed to select only those that we can utilize, that the chemical be in the original container, that a proper label be affixed, and that we are certain that the chemical is pure.

With the host of edicts from regulatory agencies, namely, EPA, OSHA, and local and state governments, chemistry departments must have established policies and procedures for the handling of their chemical waste. At the University of Akron waste solvents are handled in the following manner. Each research laboratory is issued a 5-gallon approved safety can that is kept in the solvent safety cabinet. A tag is wired to the handle of the safety can. Each time a person pours waste solvent into the safety can, the tag is updated with the person's name, the type of solvent, the quantity, and the date. When the 5-gallon safety can is full, it is delivered to the chemical stores area, whereupon the researcher receives an empty container and a new tag. The full container is transferred into an approved 55-gallon drum (utilizing a safety funnel with flame arrester and bonding wires). The information from the tag for the 5-gallon container is transferred to the 55-gallon drum manifest. In the case of very expensive solvents, some of the research laboratories choose to redistill their waste solvent. An 80% recovery rate has been reported and the redistilled products are often purer than the original start-up material.

Chemicals other than waste solvents that require disposal are brought to the chemical stores area. The chemical names and quantities are recorded on a manifest. Storage takes place in a special holding room within the chemical stores complex. According to law, the department cannot store waste chemicals for more than 90 days without a special permit from the EPA as a storage facility. Therefore, a minimum of four shipments per year are made to an EPA-approved disposal agent. Waste solvents are shipped in approved 55-gallon drums (bung type). Other chemicals must be separated into appropriate categories (e.g, Organics, Inorganics, Active

Metals) and packed in layers of absorbent packing material in separate open-head 55-gallon drums. Each drum requires the appropriate labels as mandated by the EPA and Department of Transportation. A shipping manifest that indicates the identity and quantity of each chemical accompanies each drum.

The following are some helpful recommendations regarding chemical waste:

1. Exercise a great deal of caution with regard to the type of experiments chosen for teaching laboratories. Know in advance what kind of waste will be generated. Careful thought in the selection of a laboratory experiment may prevent costly and time-consuming disposal problems later on.

2. Establish a written policy regarding chemical waste disposal and insist that it is followed. Accurate records are mandatory.

3. Be aware of the fact that there are ways to dispose of much of the chemical waste generated without the necessity of shipping it to an approved disposal agent [see *"Hazardous Chemical Waste and the Impact of R.C.R.A.," J. Chem. Ed., 331–333* (April 1982].

The author's purpose of describing the above details lies in the highly complex nature of hazardous waste disposal since the advent of the EPA regulations published in late 1980. The impact on the economic and logistical operation of all chemistry departments has been great. In designing new chemical storage facilities, or in the renovation of existing ones, it is crucial that these factors be kept in mind. Proper storage and handling facilities for hazardous waste must be well thought out and included in the overall storage scheme.

In the Knight Chemistry Laboratory at the University of Akron, the chemical stores are contained in a one-story wing separately connected to the main chemistry building. Immediate access to a freight elevator and loading dock is provided. The wing is landscaped on the three exposed sides with an earthen bank that extends to within 6 feet of the roof line. The roof itself is equipped with ventilation fans and pressure-releasing plastic blow-out panels (Figure 7.6).

Inside the building, space is divided into a receiving room, a waste solvent holding area, and three separate storage rooms for dry chemicals,

FIGURE 7.6 The main chemical storage facility. The building is earth-banked on three sides, and the roof contains pressure-releasing plastic blow-out panels.

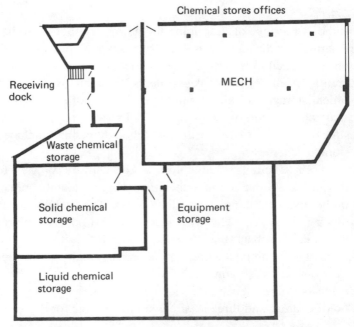

FIGURE 7.7 Design layout of the chemical storage wing.

FIGURE 7.8 An annunciator system that uses a lighted panel to pinpoint location of fire while sounding an audible alarm.

WHEN ALARM SOUNDS
VACATE AT ONCE
CARBON DIOXIDE
BEING RELEASED

FIGURE 7.9 Static-free switches and electrical receptacles reduce fire danger.

liquid chemicals, and chemical equipment, including glassware (Figure 7.7). Within the chemical storage rooms, materials are grouped generally alphabetically but overriding consideration is given to compatibility.

If a fire should break out in any of the storage areas, the smoke and heat detectors will trigger an alarm, and 30 seconds later the affected area will be blanketed with carbon dioxide. At the same time, an annunciator system located in the chemical stores office will sound an alarm, and, by means of a lighted panel, show the exact location of the fire (Figure 7.8).

To reduce still further the fire danger, all phones, switches, receptacles, and light fixtures in the chemical stores area are static-free (Figure 7.9).

7.5   CONCLUSION

All too often, the provision of adequate chemical storage space is given little consideration by university administrators when viewed as a "nonproductive" aspect of the operation. However, the lack of sufficient storage space can create hazards due to overcrowding, storage of incompatible chemicals together, and poor housekeeping. Adequate, properly designed, maintained, and well-ventilated chemical storage facilities must be provided to ensure personnel safety and property protection.

Those involved in chemical safety have a tremendous responsibility to advocate safe chemical storage facilities. University administrators and top management in industry must be convinced that chemical storerooms, although costly when done properly, are critical to the safe operation of a modern chemistry facility. To inculcate decision makers that storerooms are not to be added later as an afterthought or construed as a dead space in which shelves are erected is only part of the task.

There is a saying among architects that "design follows purpose." This is especially true when designing safe chemical storage facilities. No amount of brick and mortar, regardless of how arranged, will make an operation completely safe unless a total program of safety is present. Safe storage facilities are merely one ingredient in the total picture.

Although the safety designed into a building is vitally important, in the final analysis, people will be the determining factor as to the overall safety of the operation. People have within themselves the potential to either prevent or cause an accident. Chemical safety involves much more than sophisticated exhaust systems and careful design and layout of laboratories and storage areas, as well as the many facets of structural detail that go into designing a building. People must be taught how to be safe.

A serious and dedicated commitment to safety, with expenditure of time, energy, and money can provide a safe environment in which to work, teach, and learn.

CHAPTER **8**

IMPLEMENTATION OF AN ONLINE IMS DATABASE SYSTEM FOR WAREHOUSE AND INVENTORY MANAGEMENT

E. LAMAR HOUSTON,
University of Georgia, Athens, Georgia

8.1   INTRODUCTION

The University of Georgia organized a facility called *Central Research Stores* (CRS) during 1967. The purpose of this facility was to save the University money through volume purchases as well as to have a ready source of supply of scientific apparatus and chemicals.

Construction was finished on a 20,000-square-foot building during the fall of 1969, and shortly thereafter the operation was in full swing. All administrative functions were executed manually for 9 years, and after this period of time sales had passed the million dollar mark. The operation was beginning to drown in paperwork. The need for a computer was evident. Direct-access on-line capability was desired, so the CRS staff began talking and visiting with commercial firms to locate a software package. It was discovered that there were no inventory software packages available because of either (a) price or (b) companies did not want to give out proprietary information; therefore, university analysts and programmers began developing in-house software.

Manual operation ceased and computer operation began on March 19, 1979. After considerable transition difficulty, the computer operation smoothed out during the summer of 1979, and today it is functioning in an impressive manner. Some development work still remains, and periodically changes are necessary to some of the programs to provide more flexibility.

8.2   GENERAL SYSTEM GOALS

At the original outset of planning for the new CRS IMS database system, there were several major operating components of the CRS facility to be automated by this new system. These components were:

1. An inventory management system that would provide timely information about the inventory contained in the warehouse of the CRS facility.

2. An implementation of the basic purchasing function, including purchase order construction, issuing, and following through on the status of purchase orders once they have been placed with vendors.

3. A more extended sales function, including the order entry process, as well as the commitment of stock from the warehouse, and tracking of goods as they progress through the various steps of the sales process.

4. An interface to the then-existing manual accounting system of CRS, with an eventual progression to an automated accounting database application.

5. An easy interface to the University of Georgia accounting procedures by means of automated analysis of both the purchasing and sales functions in the CRS system.

It was also decided that the tools for implementation for this particular application would be IBM's Information Management System (IMS). Also, the system was to be an online and real-time application so that stock information displayed on terminals would reflect actual realtime stock balances within the warehouse. The large mainframe at the University's Computer Center and the IMS database system would be used with remote inquiry, update, and printing at the CRS location. This would take advantage of the staff already available and eliminate the addition of data processing staff at CRS in the event that an in-house minicomputer had been selected. And finally it was decided that the system should be recoverable and secure, both of which attributes are adequately provided by the IMS software package.

The boxes shown in Figure 8.1 represent a map of the structure of the

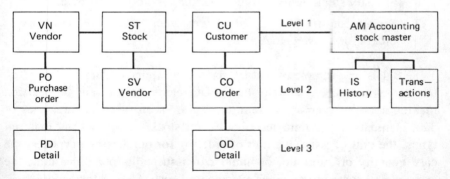

FIGURE 8.1 A map of the structure of Central Research Store Databases for the University of Georgia, Athens, Georgia.

CRS databases. The databases are all hooked together in a logical manner that allows browsing from one database to another. All one must do is key in the segment to reach a specific database. For example, in order to move from the vendor database (VN) to the customer database (CU), one must key in the segment CU. One can also reach different levels of each database by keying in the various segments listed on the map.

8.3   INVENTORY PROCESSING AND INFORMATION

The original goals for this particular phase of the application were as follows:

1. Maintain a stock master with information about specific stock items, including the description, issue units, purchase units, quantity on order, quantity on hand, quantity backordered, and other various statistics about each item.

2. Automatically maintain receipts and issue information about each specific stock item.

3. Provide varied lists for taking inventory and analysis of stock items.

4. Generate an automated sales catalog with camera-ready proof.

5. Generate any miscellaneous stock-related reports as required on an ad hoc basis.

6. Classify stock items by type, markup, and other attributes.

7. Provide on-line inquiry of all customer orders and purchase orders for a given stock item.

Table 8.1 is an example of a stock database inquiry update screen. Contained on this screen are the basic data elements relevant to decision making about a given stock item. As can be seen, various categories of data elements are maintained about each stock item, such as material types, the unit of issue, the unit of ordering for purchasing purposes, the case quantity or multiplier of number of issue units per order unit, the average unit cost, and previous average unit cost. Also, information about stock availability is readily available by looking at this screen. History

TABLE 8.1 Stock Database Inquiry/Update Screen

```
                STOCK DATABASE INQUIRY/UPDATE                    08/16/82
                                                                15:21:35
STOCK NUMBER    102900      ST COMMODITY  2899735
DESC            ACETONE AR CH32CO         FLAMMABLE 209

MATERIAL TYPE   11              CHEMICAL EXP        COST CODE     2
STOCK TYPE      1               REGULAR             UNIT COST          7.2200
ISSUE UNIT      4L              CASE QTY        1   FREV COST          7.3569
ACTIVE CODE                     ORDER UNIT     4L
ON HAND            13           SUSPENSE         0   15     %         8.3080
COMMITTED           0           AVAILABLE       13   BOOKSTORE        9.2317
ON ORDER            0           BACK ORDER       0
LAST ISSUE      08/09/82        LAST RECEIPT  06/29/82
PERIOD ISSUES     493           YEAR TO DATE     493
RECEIPTS          442                            442
DEMAND            493                            493
STOCK OUTS          4                              4
BACKORDERS         27                             27
OPTION:         TRX: 5ST    STOCK NUMBER: 102900
   *** ENTER DATA FOR UPDATE ***
```

information about period issues and receipts, demand, stockouts, and backorders are also readily accessible.

The previously discussed screen, as well as many others, was implemented using IBM's IMS software package as well as their Application Development Facility (ADF) package. The ADF package has proved to be a very effective productivity enhancer at the University of Georgia, allowing online IMS applications to be implemented in one-fourth to one-twentieth the time of a pure COBOL or PL/1 IMS application. By using ADF for online inquiry and update in conjunction with Informatic's MARKIV for batch processing, significant productivity enhancements have been obtained over the past several years, allowing rapid implementation of systems similar to this particular application.

Other screens have been developed but are not displayed in this chapter, which allow a CRS staff member to inquire into all of the outstanding customer orders by customer number, as well as purchase orders by vendor number outstanding for a given stock item. Along with this information are provided various status codes to indicate the status of each particular customer or purchase order in its normal life span. An additional screen was also provided to show all vendors who provide a given stock item, along with the representative for that vendor, the vendor catalog number, and the telephone and address of the vendor. This information can be used in constructing purchase orders.

Currently the normal "reorder at minimum, reorder up to maximum" procedures are being used for replenishing stock at CRS; however, when the original CRS system was implemented, the data elements were provided in the stock segment of the database to be able to easily interface with IBM's INFOREM package. This package is a very sophisticated inventory forecasting and management package that has been purchased by CRS for later installation with the IMS inventory management system, which is the topic of this chapter. The data elements for each stock item are already stored in the database. The basic functions of INFOREM, when completely installed, will be to (a) monitor and measure the average current demand for items at the stock-keeping unit level, (b) produce forecasts of future stock-keeping unit demand, (c) develop ordering decision rules for optimum restocking of items, (d) maintain accurate lead time for vendors at the item level, and (e) simulate performance for a month, season, or year.

8.4　PURCHASING

The original goals for the purchasing component of this system were to provide (a) real-time inquiry and update about vendors and purchase orders for each vendor, (b) vendor cross-reference lists, and (c) real-time information about all orders for a stock item and also to (d) control the status of a purchase order at the detail line level—in other words, to be able to accommodate the partial completion, partial payment, and partial receipt of specific lines or parts of lines on a given purchase order.

Several screens were developed to fulfill the on-line inquiry requirement of this component. A vendor database inquiry update screen was designed to show the following: the vendor number and name, as well as the status code for the vendor; payment terms required, shipping terms, any outstanding contracts that might exist for a vendor; the vendor representative and phone number; and purchase from and pay to addresses. By using ADF's automatic segment browsing capability, a CRS staff member can easily browse through all outstanding purchase orders for a given vendor. Also, by proceeding to the second level of the database, this staff member can also browse through all the detail lines for a given purchase order, showing quantities ordered and received, the unit price, vendor reference number, and the particular stock number associated with the detail line (Table 8.2). At the detail, or line, level of a purchase order

TABLE 8.2   Vendor Database—Purchase Order Detail

```
                    VENDOR DATABASE - PURCHASE ORDER DETAIL         08/16/82
                                                                    16:12:11
PURCHASE ORDER      CRS-4213-A
LINE/DASH           002        01              STATUS    1
STOCK ITEM          2211002001                PUR/ISS INDIC   P
DESCRIPTION         A-540 ALUMINA 80-200MESH   SPECIAL ORDER ITEM

QUANTITY ORDERED          1                   QUANTITY RECEIVED          1
ORDER UNIT PRICE         42.64                PURCH UNIT PRICE          42.64
VENDOR PACK SLIP    019885                    VENDOR INV REF NO
VENDOR PYMT REF#                              ORDER DATE          07/30/82
                    --------- STATUS --------      -- DATE --
RECEIVING                1         RECEIVED         08/03/82
INVCD BY VENDOR          1         INVOICED         08/13/82
PAID TO VENDOR           0         NOT PAID           /  /
OPTION:           TRX:   5PD
NEXT PO DETAIL NUMBER:   04700CRS-4213-A00201
 *** ENTER DATA FOR UPDATE ***


            S E C O N D A R Y   K E Y   S E L E C T I O N
    UPDATE            TRANSACTION: VENDOR
    OPTION:      TRX: 5PD    KEY: 04700CRS-4213-A00000
    SELECTION:          *** ENTER A SELECTION NUMBER FROM THIS SCREEN ***
                        QUANTITY    QUANTITY      UNIT    VENDOR     STOCK
             LINE  CHNG ORDERED     RECEIVED      PRICE   REFERENCE  NUMBER
       1     001   01      1           1          47.00             2208026001
       2     002   01      1           1          42.64             2211002001
       3     003   01      1           1          81.00             2211003002
       4     004   01      1           0          43.88             2211004001
       5     005   01      1           1          26.33             2211004002
       6     006   01      2           2          21.06             2211005001
       7     007   01      1           1          23.08             2211006001
       8     008   01      1           1          30.42             2211007001
       9     009   01      1           1          25.25             2211011001
      10     010   01      1           1          48.41             2211011002
      11     011   01      2           2          23.12             2211013001
      12     012   01      2           0          10.47             2211013002
      13     013   01      1           1          13.50             2211013003
      14     014   01      6           6           1.52             2211016001
      15     015   01      3           3          11.82             2211016005
```

(Table 8.2) data elements are carried to show the actual item ordered from the vendor, as well as the quantity ordered and received, the order unit price and purchase unit price, the vendor's packing slip number, vendor payment reference numbers, and the original order date. The various statuses of a given detail line, that is, the receiving, invoicing, and payment status codes, are carried at the detail line level reflecting whether each of these particular stages has been reached and the date on which the stage was reached for that line. Several screens have been provided to allow the CRS staff to inquire into a given purchase order and examine the status codes thereof.

Figure 8.2 shows the general flow of purchase order construction. The normal steps involved in this process are:

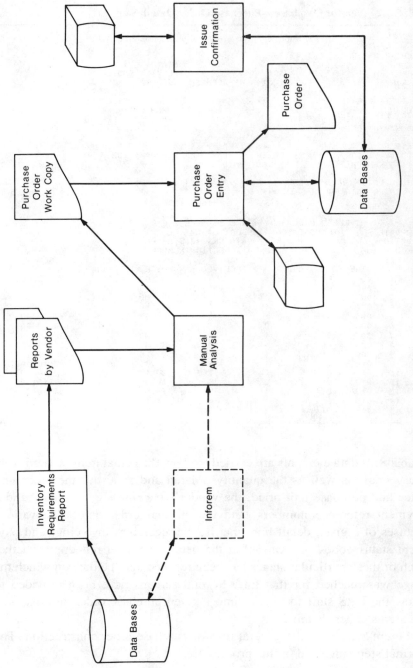

FIGURE 8.2 General flow of purchase order construction.

1. Determine the items to be ordered.

2. Obtain prices for items.

3. Select vendors for the particular items.

4. Enter the purchase orders into the system through the CRT.

5. Print purchase orders on the printer.

6. Confirm the purchase order as being issued.

Up to the point in time where the purchase order is actually confirmed as issued, the CRS staff member can modify the purchase order contents in the database as necessary. However, after purchase order issue confirmation is done, all changes to the purchase order are treated like change orders, and consequently account effects are generated in the accounting reports to reflect these as change orders or purchase order adjustments.

To ensure that the purchasing staff knows what items to reorder, a "Stock Items Below Minimum Level" report is generated showing by vendor the necessary stock items required of that vendor, as well as quantities required to refresh the stock level in the warehouse. The actual entry of the purchase order involves the calling up of a blank screen and entering of the vendor number, purchase order number, and, for each line on the purchase order, the stock number, quantity ordered, and the price at which the quantity is being ordered. The person entering this purchase order has the option of taking the default purchase unit or specifying that the stock item is being ordered by issue units. This ability to specify that a given line on a purchase order is either in purchase or issue units allows the receiving of partial damaged cases at receipt time. On entry of the screen with the previously mentioned data elements, the purchase order is stored in the vendor purchase order database and a confirmation is returned to the screen. This confirmation also contains the purchase unit and the various status codes for the purchase order (Table 8.3).

By using the purchase and check request attachment screen (Table 8.4), the CRS staff member can then request that the purchase order that had just been entered be printed on the remote 3287 printer. On this purchase order printing screen, the vendor number and purchase order are entered along with selection criteria: whether only received or not received items, invoiced or not invoiced items, and paid or not paid items should be printed. Also, the option of whether to extend prices is provided, as well as the ability to specify an alternate printer. When the purchase order is

TABLE 8.3 Confirmation Screen for Vendor Purchase Order

```
VENDOR 04700    PO NUM CRS-4213-A  FUNC    REF#                        82/08/16
OK
LN   CH   STOCK       QTY     UNIT      COST    FREIGHT  I/P    RECV   S  R  I  P
001  01  2208026001    1     EA        47.00                    P       1   1  1  1  0
002  01  2211002001    1     CS/6      42.64                    P       1   1  1  1  0
003  01  2211003002    1     EA        81.00                    P       1   1  1  1  0
004  01  2211004001    1     EA        43.88                    P       0   1  0  0  0
005  01  2211004002    1     PK/M      26.33                    P       1   1  1  1  0
006  01  2211005001    2     PK/12     21.06                    P       2   1  1  1  0
007  01  2211006001    1     100GM     23.08                    P       1   1  1  1  0
008  01  2211007001    1     CS/4      30.42                    P       1   1  1  0  0
009  01  2211011001    1     PK/6      25.25                    P       1   1  1  1  0
010  01  2211011002    1     100GM     48.41                    P       1   1  1  1  0
011  01  2211013001    2     CS/500    23.12                    P       2   1  1  1  0
012  01  2211013002    2     CS/144    10.47                    P       0   1  0  0  0
013  01  2211013003    1     EA        13.50                    P       1   1  1  1  0
014  01  2211016001    6     EA         1.52                    P       6   1  1  1  0
015  01  2211016005    3     EA        11.82                    P       3   1  1  1  0
```

printed (Table 8.5), it is printed on 8½ × 11-inch white unlined paper and is in a suitable form for attachment directly to the "University Purchase and Check Request" form. This is sent directly to the vendor.

The receiving of goods consists of the following major steps:

1. Recording the receipt of orders in the computer system.

2. The recording of confirmation of payments to the vendor by the University of Georgia's Accounts Payable Department.

3. The provision of reports at each of these preceding stages.

4. Accommodation of change orders, either before or after receipt of goods from the vendor.

TABLE 8.4 Purchase and Check Request Attachment Screen

```
                    C E N T R A L   R E S E A R C H   S T O R E S
                    PURCHASE AND CHECK REQUEST ATTACHMENT
         DATE              08/16/82
         VENDOR NO         04700
         PURORD NO         CRS-4213-A
         VPSL NO
                                    ENTER X BELOW TO SELECT BOTH OPTIONS
             RECEIVED        1      0=NOT RECV       1=RECV
             INVOICED        0      0=NOT INVOICED   1=INVOICED
             PAID            0      0=NOT PAID       1=PAID
                             Y      EXTEND PRICES    Y OR N
                             CRSVP1 PRINTER - USUALLY CRSVP1
```

TABLE 8.5   Computer-Printed Purchase Order

THE UNIVERSITY OF GEORGIA			REQUEST NO	CRS-4213-A	
PURCHASE AND CHECK REQUEST ATTACHMENT			(04700)		
			PAGE NO	1 OF 1 PAGES	

ITEM NO	DESCRIPTION & SPECIFICATIONS	QUANTITY UNIT	INTERNAL USE	UNIT PRICE	TOTAL PRICE
001-01	2208026001 13-761-14A PLATINUM FOIL	1 X EA		47.00	47.00
002-01	2211002001 A-540 ALUMINA 80-200MESH	1 X CS/6		42.64	42.64
003-01	2211003002 13-681-15 PIPET AID FILLE	1 X EA		81.00	81.00
005-01	2211004002 21-341 MLA PIPET TIPS	1 X PK/M		26.33	26.33
006-01	2211005001 09-740-5 MEMBRANE FILTER	2 X PK/12		21.06	42.12
007-01	2211006001 1005 ETHYLENE OXIDE	1 X 100GM		23.08	23.08
009-01	2211011001 09-730-225 FILTER HOLDER	1 X PK/6		25.25	25.25
010-01	2211011002 14707 GLUTARALDEHYDE	1 X 100GM		48.41	48.41
011-01	2211013001 05-407-5 CENT TUBE MICRO	2 X CS/50G		23.12	46.24
013-01	2211013003 21-344 PIPET RACK	1 X EA		13.50	13.50
014-01	2211016001 11-875-50 TAPE INDICATOR	6 X EA		1.52	9.12
015-01	2211016005 13-681-51 PIPET FILLER	3 X EA		11.82	35.46
	PO TOTAL =				440.15

5. The allowance of other adjustments to purchase orders.
6. The handling of freight charges, which are added on to a specific purchase order and the redistribution of these charges, if necessary, back to the original ordering customer.

Figure 8.3 shows how receiving reports and changes in amendments to orders are entered through the terminal into the vendor/purchase order database. Various vendor analysis reports are provided from this database,

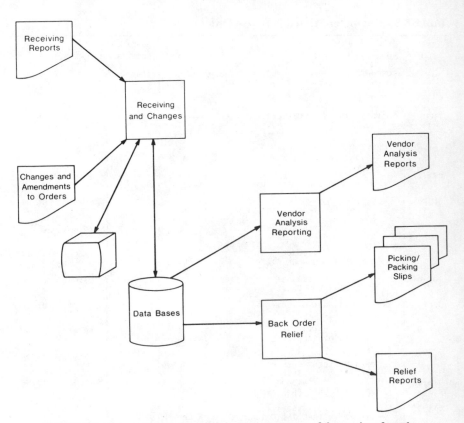

FIGURE 8.3 Schematic flow chart for the computer entry of the receipt of goods.

and automatic backorder relief programs are run overnight to scan these databases and automatically print picking and packing slips the following morning to relieve backorders. In addition to various backorder relief reports, this feature eliminates the need for constant examination by the CRS staff of receiving reports to see which backorders can be relieved.

8.5 SALES PROCESSING

The customer order entry phase of this system includes the following major functions:

1. Recording the original customer orders.

2. Printing picking and packing slips to be sent to the warehouse.

3. Adjustment of inventory and customer orders, as necessary, based on reconciliation with the actual stock in the warehouse.

4. Shipping the goods to the customer by freezing the price in the database, and so on.

5. The automatic generation of an invoice to be sent to the customer at billing time.

6. The handling of multiple generations of a customer order as previously nonavailable items are received through backorder processing.

Figure 8.4 shows the normal flow of the customer sales function. Written orders or telephone calls are entered through the terminal to the customer database, picking and packing slips are sent to the warehouse, and then on confirmation that the goods have been shipped or delivered, the database is updated through the terminal and an invoice is automatically printed for filing by customer number.

Several screens were provided for inquiry into the availability and prices of several stock items at one time (Table 8.6). This allows immediate access to this information in response to telephone inquiries or written orders. The customer order entry screen can also be used for inquiring into the status of a specific customer order (Table 8.7). A customer order is actually entered by calling up a blank customer order entry screen and entering the customer number, order number, a ship to address if different from the normal billing address, and then the detail lines for the customer order. Each line requires only a stock number and a quantity ordered. The computer system automatically checks stock available and splits lines that are partially available into items to be shipped and items to be back-ordered. The average unit price, currently available in the stock database, is frozen for items to be shipped and extended prices are automatically calculated.

The sales clerk then can use a picking/packing slip invoice printing screen (Table 8.8) to enter the customer number, customer order number, and request any given generation picking/packing slip for a given customer order. Also on this screen, the options previously mentioned for the pur-

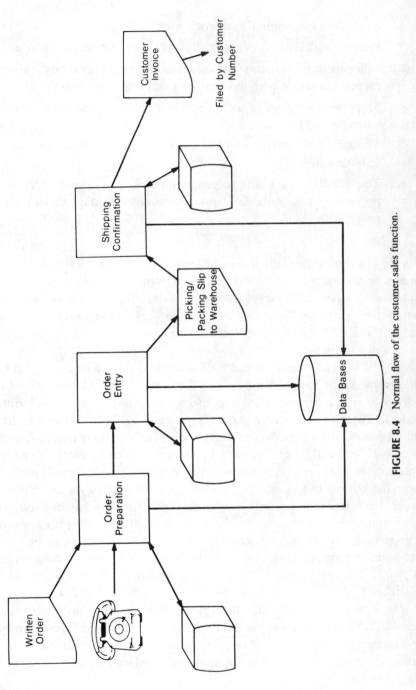

FIGURE 8.4 Normal flow of the customer sales function.

TABLE 8.6 Integration of Computer Screens for Stock Availability

```
102900                  4L        I 106110                  4L
   ACETONE AR CH32CO               I    ACETONITRILE HPLC
                                   I
   UNIVERSITY PRICE  $     8.30    I    UNIVERSITY PRICE  $      22.05
   BOOKSTORE  PRICE  $     9.21    I    BOOKSTORE  PRICE  $      24.48
   COST              $     7.22    I    COST              $      19.17
   AVAILABLE QTY ====>      13     I    AVAILABLE QTY ====>       10
===================================I ===================================
110500                  4 LB      I 191500                  500ML
   ALCONOX                         I    BUFFER SOLUTION PH7
                                   I    YELLOW
   UNIVERSITY PRICE  $     5.14    I    UNIVERSITY PRICE  $       3.28
   BOOKSTORE  PRICE  $     5.71    I    BOOKSTORE  PRICE  $       3.64
   COST              $     4.47    I    COST              $       2.85
   AVAILABLE QTY ====>      85     I    AVAILABLE QTY ====>       95
===================================I ===================================
164820                  4L        I 234300                  4L
   BENZENE AR C6H6                 I    CHLOROFORM AR
                                   I    CHCL3
   UNIVERSITY PRICE  $    11.43    I    UNIVERSITY PRICE  $      18.33
   BOOKSTORE  PRICE  $    12.69    I    BOOKSTORE  PRICE  $      20.35
   COST              $     9.94    I    COST              $      15.94
   AVAILABLE QTY ====>       0     I    AVAILABLE QTY ====>       20
```

TABLE 8.7 Customer Order Entry Screen

```
CUSTOMER 00024   ORDER 2211002   FUNC    SHIP TO ROBERT SMITH
REF              DASH             CONF            CHEMISTRY
        82/08/16                                 CHEMISTRY RM. 101
MORE
```

LINE	DASH	STOCK	ORDER	BO	I SHIP	UNIT PRICE	S	I	P	DASH
001	01	2211002001	1		1	49.04	2	1	0	002
002	01	164800	12		12	2.38	2	1	0	001
003	01	164820	6		6	11.43	2	1	0	001
003	02	164820		6	0	11.43	3	0	0	000
004	01	235900	10		10	18.19	2	1	0	001
004	02	235900	2		0	20.16	·	0	0	003
005	01	472100	12		12	4.49	2	1	0	001
006	01	472500	10		10	10.23	2	1	0	001
006	02	472500	2		0	11.20	1	0	0	003
007	01	673205	24		24	1.76	2	1	0	001
008	01	714500	4		4	6.39	2	1	0	001
009	01	714800	4		4	8.60	2	1	0	001
010	01	782730	12		12	9.46	2	1	0	001
011	01	784700	12		12	6.92	2	1	0	001
012	01	794900	21		21	4.97	2	1	0	001
012	02	794900		3	0	4.97	3	0	0	000
013	01	848300	180		180	0.92	2	1	0	001
014	01	947600	1		1	7.44	2	1	0	001

TABLE 8.8   Picking/Packing Slip—Invoice Printer Screen

```
                    C E N T R A L    R E S E A R C H    S T O R E S
                    PICKING/PACKING SLIP - INVOICE PRINTER
DATE                08/16/82
CUSTOMER NO         00024
ORDER NO            2211002
PICKING SLIP#:      003
   PICKED/SHIPPED   1          0=NOT PP    1=PP NOT SHPD    2=SHIPPED    3=BACKORD
   INVOICED         X          0=NOT INV   1=REG INV        2=PRO FORMA
   PAID FOR         X          0=NOT PAID  1=PAID
EXTEND PRICES?      N          N Y I OR P
PRINTER TERMINAL    CRSVP2     USUALLY CRSVP2
RESPONSE
```

chase order printing screen are available. That is, the clerk has the option of just getting a normal picking/packing slip or selectively printing a customer order showing only lines that have been picked and packed but not shipped, or only the lines that have been shipped, or only the lines that have been backordered. Also, by selecting a different status code, items either invoiced or not invoiced can be printed. Other options are whether to extend prices and the specification of an alternate printer.

The customer billing function has four major subfunctions:

1. The recording of nonuniversity payments and associated control information.

2. The automatic collection or reimbursal, if necessary, to university customers through the university accounting system.

3. The printing of customer statements on demand with ability to freeze orders at cutoff for billing processing. In other words, the system can continue to receive orders from customers without affecting what was shown on the statements as they were originally at the end of the billing period. Rerunning of statements will produce the same results because of the date effective logic used in implementation of this function.

4. Controlling the status of customer orders at a detail line level, yet allowing the entry of commands that affect all of these lines at an aggregate level. In other words, given lines can be ordered, received, invoiced, and paid, yet the CRS staff does not have to enter commands at the detail line to have this function performed.

Figure 8.5 shows the normal processing of the customer billing function. The databases are scanned and statements are printed. Automatic journal vouchers are generated to go to the university accounting office, as well as automated transactions to go into the university's property control system for equipment items that were purchased through CRS. The statements are taken and analyzed. Reconciliation occurs by comparing the originally printed invoices and making sure that the information in the invoices matches that on the statements. If necessary, statements can be reprinted. On the final completion of statement printing, the invoices are bundled with the statements and shipped to the customers. No action is required of university customers at this point. The funds are collected automatically from their expense accounts. For nonuniversity customers, however, payments are received and are entered through the terminal to indicate that the specific lines or entire orders for given customer orders have been paid.

At the detail customer order line level in the customer database (Table 8.9), data elements are carried to indicate what the specific order line represents as well as its status. Descriptive data elements include the stock number, quantity ordered, quantity shipped, the frozen unit price for this specific order line, the unit cost to CRS, and payment descriptive information (e.g., the journal entry or ticket number for a university payment, if appropriate). The status codes, as in the purchasing function, are the shipping, invoicing, and payment status code showing whether the specific line has been shipped, invoiced, and/or paid, and the specific dates on which these stages were reached.

Several special processes and reports were also provided in sales processing. For example, at the end of the fiscal year, for items that have been back-ordered but are still not received for customers, the system has the ability of cutting a "pro forma" invoice. These invoices allow the customers to be invoiced before the end of the fiscal year and thus have their expense account charged during the current fiscal year. When the items are received in the subsequent fiscal year, they are already handled as having been invoiced. A pro forma status report is provided to all customers to show those items still in a pro forma invoice status. Another special process is the handling of walk-in customers. A walk-in customer has an additional markup for handling of the specific order; therefore, the pricing for walk-in customers is different. Also, sales tax must be handled for walk-in customers. As previously mentioned, backorder relief is auto-

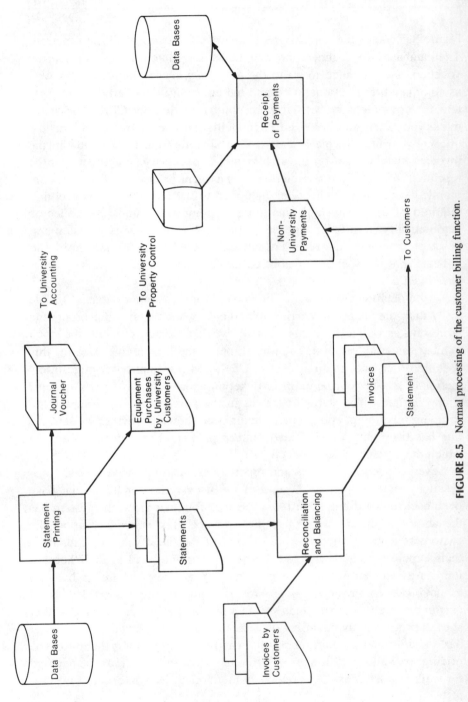

FIGURE 8.5 Normal processing of the customer billing function.

190

TABLE 8.9 Detail of Computer Screen of Customer Database

```
                CUSTOMER DATABASE - ORDER DETAIL                   08/16/82
                                                                   15:38:54
LINE NUMBER    001            DASH NUMBER   01         PP SLIP NO  002
STOCK NUMBER           2211002001
QUANTITY ORDERED            1
QUANTITY SHIPPED            1
UNIT PRICE             49.04
UNIT COST              42.6400
PAYMENT NUMBER
                       ---------- STATUS ----------      -- DATE --
SHIPMENT               2           SHIPPED               08/04/82
INVOICING              1           INVOICED              08/04/82
PAYMENT                0           NOT PAID               /  /
DESCRIPTION            A-540 ALUMINA 80-200MESH    SPECIAL ORDER ITEM

OPTION:        TRX: 50D
NEXT LINE-DASH NUMBER:  00024221100200101
   *** ENTER DATA FOR UPDATE ***
```

matically performed by the system. Miscellaneous sales reports are provided, as well as various customer cross-reference lists.

8.6 ACCOUNTING INTERFACE

As originally stated, the goals for this system were to provide an easy interface to the then existing CRS manual accounting books. Because of this, the original functions as implemented were the generation of transactions on a daily or other basis by scanning the customer and vendor databases, generating daily journals of transactions, and posting these manually to the CRS books. By adding other transactions to the types of transactions generated originally, a full-scale accounting system has subsequently been developed. These transactions go directly into an accounting database, and a full-scale "general ledger budgetary accounting system" has been implemented that carries online detail information about all transactions against specific accounts. The types of transaction handled are:

Journal entries	Invoices
Walk-in sales	Pro forma invoices
Cash receipts	Monthly billing
Purchase orders	Purchase order adjustments
Receiving reports	Vendor invoices
Completion payments	Retroactive adjustments

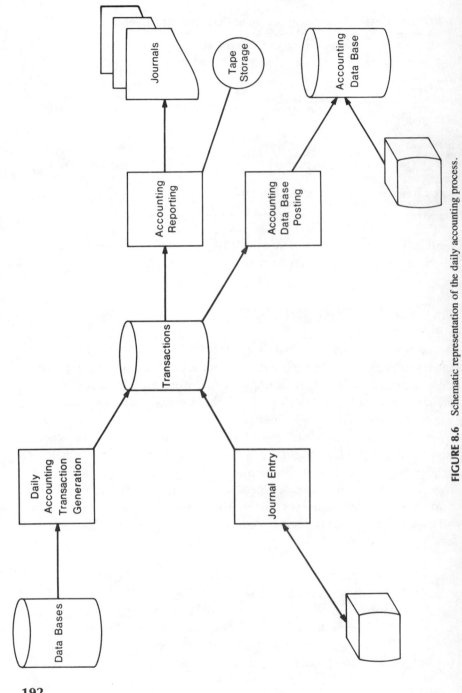

FIGURE 8.6 Schematic representation of the daily accounting process.

TABLE 8.10   Journal Entry Screen

```
                        C R S   JOURNAL ENTRY SCREEN              82/08/16
JE NO: 747             DATE: 820806 (YYMMDD)
REF#1: 07956           REF#2: 5961450       REF#3: CK#517472
TOTAL DEBIT = CREDIT
ACCT          AMOUNT      D/C      ACCT DESC
211           225.00      D        ACCOUNTS PAYABLE
1121          225.00      C        RESERVE STOCK FUNDS




EXPLAINATION:
TO RECORD PAYMENT FROM THE 7/31/82 STATUS REPORT FOR JUNE TANK RENTAL FOR
LIQUID NITROGEN VPS#019098.
```

Figure 8.6 shows the daily accounting process, which is initiated by the scanning of databases and generation of daily accounting transactions. Journal entries are entered directly through a journal entry screen (Table 8.10), at which time a debit/credit balancing and an account validation function is performed. These journal entries are then added to the daily accounting transactions produced by scanning the databases, and various accounting reports are provided. Account transactions are then stored on tape storage as well as posted to an accounting database for subsequent on-line inquiry. The transactions generated by scanning the databases have predetermined general ledger effects. These general ledger effects are easily changed by modifications of tabulated entries in the system.

Table 8.11 shows an account master (AM) screen for one account, cash

TABLE 8.11   Account Master Screen

```
VDRS10AC              ACCOUNTING DATABASE INQUIRY            08/16/82
                                                            18:25:48
OPTION:        TRX: 5AM
KEY   : 1111
*** ENTER DATA FOR UPDATE ***
ACCOUNT NUMBER: 1111
DESCRIPTION:   CASH ON HAND
BUDGT:         .00        ENCUMB:          .00      FREE BAL:    13729.36
                                    BALANCES
                DEBITS            CREDITS        NET          TYPE
LAST POSTING           .00              .00             .00    D
LAST MONTH        145767.23       145341.87          425.36    D
MONTH TO DATE       3602.91             8.33         3594.58   D
LAST YEAR        2027287.85      2017578.43         9705.42    D
YR-TO-MONTH BEG   145767.23       145341.87          425.36    D
YEAR TO DATE      149370.14       145350.20         13729.36   D
```

TABLE 8.12   Accounting History Screen

```
VDRS10AH                        ACCOUNTING HISTORY                   08/16/82
                                                                     16:29:22
OPTION:          TRX: 6IS
KEY     : 1111                  1982

ACCOUNT NO:   1982                        YEAR:   1982
DESCRIPTION:   CASH ON HAND
                          DEBITS            CREDITS
                                                   STARTING BAL      4842.97
         MONTH 1   JUL     153484.28       187171.71     -28844.46
         MONTH 2   AUG     147320.64       168149.42     -49673.24
         MONTH 3   SEP     247568.19       138427.64      59467.31
         MONTH 4   OCT     257281.72       224212.23      92536.80
         MONTH 5   NOV     130371.63       116765.39     106143.04
         MONTH 6   DEC     110961.37        97296.23     119808.18
         MONTH 7   JAN      99888.69        89280.35     130416.52
         MONTH 8   FEB     148380.54       128413.91     150383.15
         MONTH 9   MAR     126751.78       111275.51     165859.42
         MONTH 10  APR     183205.69       163379.16     185685.95
         MONTH 11  MAY     171890.95       147141.14     210435.76
         MONTH 12  JUN     245339.40       432127.71      23647.45
         CLOSING AMTS          .00          13938.03          .00
         TOTAL             2022444.88      2017578.43 ENDING BAL     23647.45
```

on hand. Amounts are rolled on a monthly basis within the account master segment as well as to the history segment (IS) (see Table 8.12) of the accounting database. Table 8.13 shows a transaction screen (TR) for listing current monthly transactions. Individual transactions are kept in this portion of the accounting database, and the amounts are rolled into the appropriate portion of the account master segment on a daily basis. When-

TABLE 8.13   Transaction Screen for Listing Current Monthly Transactions

```
          S E C O N D A R Y   K E Y   S E L E C T I O N
RETRIEVE        TRANSACTION: ACCOUNT TRANSACTIONS
OPTION: F    TRX: 6TR   KEY: 1111                   1982
SELECTION:         PRESS ENTER TO DISPLAY ADDITIONAL SELECTIONS
                              TRANS KEY
       1    8207221114         2203046001C00052              BSC
       2    8207221114         2203046001C00052              BSD
       3    8207221115         2203046001C00052              BSC
       4    8207221115         2203046001C00052              BSD
       5    820722113          2117018003C07165              IND
       6    820722113          2133031003C08280              IND
       7    820722113          2138032001C01686              IND
       8    820722113          2140041003C03962              INC
       9    820722113          2146020005C00024              INC
      10    820722113          2158004001C00317              IND
      11    820722113          2173031001C07872              IND
      12    820722113          2176002001C05934              IND
      13    820722113          2181007001C08916              IND
      14    820722113          2182009001C00283              IND
      15    820722113          2188018001C07656              IND
      16    820722113          2188018003C07656              IND
      17    820722113          2193007001C05934              IND
```

ever the current accounting month ends, individual transactions are "cleared" out, and the transaction portion of accounting database is clear and ready for a new accounting month.

8.7 CONCLUSION

In conclusion, the computer has had a profound impact on all segments of CRS operation. It has allowed a rate of change in the data processing area that would not have been possible under a manual or possibly a minicomputer system.

The annual operating cost of the computer is somewhat less than the cost of one classified employee. No employees have been added to the operation since the computer arrived, even though we have experienced an annual growth rate of 20% a year. We do not anticipate adding any employees in the near future.

The trend into the future seems to be that institutions and corporations are dramatically shifting into computerized operations either with a minicomputer or hooked to a mainframe. It must be emphasized that no system should be implemented without proper safeguards for security, recoverability, and auditability. The future ahead in the 1980s will certainly be challenging if the rate of change in the data processing area continues to grow at the rate at which it has over the past decade. It becomes almost inconceivable to visualize what administrative roles will be like for direct users and data processing professionals in the year 1990.

CHAPTER **9**

A PROVEN PLAN
FOR ELIMINATING
DANGEROUS CHEMICALS
FROM SCHOOLS

J. GERLOVICH
Iowa Department of Public Instruction, Des Moines, Iowa

9.1   INTRODUCTION

At the secondary educational level (grades 7–14), the type of experiments conducted in science classes and the selection and management of appropriate reagents constitute two of the most critical means of maintaining a safe learning environment. With increasing propensity of the public to file tort liability suits when a wrong is perceived, teachers must protect themselves by becoming better informed concerning the benefit/risk value of the chemicals selected for inclusion in their science curriculum.

According to Coble and Hounshell (1),

> The questions science teachers should be evaluating relate not only to the alleged, or known, risk of the substance, but also to:
>
> 1. The perceived importance of the educational objective that the laboratory exercise is designed to meet.
> 2. The existence (or nonexistence) of alternative ways of meeting the objective, or of chemical substitutes.
> 3. The method by which the exercise is to be conducted—whether as a hands-on student activity or as a teacher demonstration.
> 4. The maturity and/or "competence" of the students.
> 5. Environmental conditions within the laboratory (adequacy of ventilation; laboratory design, including safety features; storage facilities; etc.).

In a 1961 survey of California chemistry teachers, Macomber reported that "several teachers finished their college chemistry courses with only vague ideas about the dangers involved in certain experiments or in the use of certain chemicals" (2). In 1974 two Paxton Center (Massachusetts) School 13-year-old girls were severely burned, as a result of misinterpretation of experimental procedures outlined in their textbook (3). The total suit involved $825,000. In 1975 a California high school student was injured in a chemistry laboratory when a chemical mixture exploded in his hands. Although the student had substituted potassium chlorate for potassium nitrate (specified by the textbook), he alleged that he had not been warned of the hazard of substitution. The student filed suit alleging teacher negligence in not providing instruction pertaining to chemical substitution. The court concluded that it *is* the teacher's responsibility to personally instruct students regarding selection, mingling, and use of ingredients in potentially dangerous mixtures, including instructions concerning chemical substitutes (4).

In October 1979 a northern Iowa junior high school teacher experienced the symptoms of cardiac arrhythmia following a full teaching day conducting experiments in which nitrous oxide fumes were liberated in the laboratory. Since no exhaust hood was available, experiments were conducted on the window sill, with windows ajar. Following admission to the local hospital, the attending physician diagnosed nitrous oxide poisoning due to cumulative effects (5).

During fall 1979 some long-forgotten picric acid (potential explosive chemical under certain conditions) was discovered on the shelves of many Pennsylvania schools. The overreaction of some officials touched off a series of repercussions in many other states. McDermott and Edgar reported that the whole episode could have been avoided if science teachers had heeded the repeated warnings to clean out storerooms and discard unwanted or unused chemicals (6).

At the Tenth Biennial Education Conference of the American Chemical Society (ACS) (October 1977) the participants recommended (7):

1. . . . that it (ACS) establish an adequately funded office of chemical safety and health, and that this office. . . . Provide academic chemistry departments with information about internal procedures and outside services for the disposal of chemical wastes . . .

13. . . . to elementary and secondary schools that they give greater attention to safety and health in their science courses . . .

16. . . . to manufacturers and distributors of chemicals that they provide catalogue listings and labels that clearly indicate the hazard class of the contents.

It appears that steps are being taken nationally to raise the awareness level of science teachers relevant to chemical safety in academic settings. At least two specific questions, however, remain to be addressed:

1. Which chemicals present greater hazard than their inclusion merits in the secondary school science curriculum?

2. Considering newly adopted Environmental Protection Agency (EPA) regulations concerning handling, transport, and disposal of certain substances (many of which *are* in schools); manufacturers' inability to retrieve and dispose of these hazardous substances; pressure by insurance companies, fire marshall's offices, and school boards to purge such chemicals from storerooms

and the lack of pragmatic disposal techniques; how do local schools remove such unwanted chemicals?

Many of the details of these questions are addressed in the plan outlined below. The experience of one state indicated that there are no easy, quick answers. Through the cooperative efforts of all agencies responsible, however, the problems can be addressed safely and effectively.

9.2   IOWA EXPERIENCE WITH PICRIC ACID

9.2.1   The Tip of the Iceberg

Picric acid was originally used in science classes to stain cell organelles for microscopy work. As textbook science programs changed, better staining techniques and chemicals were introduced to teachers. Picric acid slowly fell into disuse. Over the years, without regular inventories and/or purging, the acid accumulated in the storerooms of many schools. As the chemical dried out, it became increasingly unstable.

In early April 1979, an article appeared prematurely in the Iowa State Superintendent's Newsletter, *Outreach,* entitled "Picric Acid Means Danger in School Lab" (8). The newsletter, which is mailed regularly to all Iowa educational institutions, created a series of repercussions. Many principals, superintendents, and classroom teachers immediately requested assistance in disposing of the acid. At that time, no convenient, effective mechanism existed to provide safe, practical, inexpensive, and expeditious disposal of the potential explosive. Manufacturers could not or would not retrieve such small quantities of these chemicals because of transportation regulations, containers missing lot numbers, and diverse geographic distribution. In addition, the total magnitude of the problem, including quantity, physical state, and exact location of the substances was unknown.

In late April 1979, a chemical inventory entitled *Assessment Inventory of Picric Acid in Iowa Schools* (Table 9.1) was sent by the science consultant of the Iowa Department of Public Instruction (DPI) to all secondary schools (grades 7–12) and community colleges to ascertain the magnitude of the problem. Teachers were instructed to note all forms of the picric acid in their inventories, including $C_6H_3N_3O_7$, ammonium picrate, 2,4,6-

TABLE 9.1   Inventory Sheet Used to Assess Amounts of Picric Acid in Iowa Schools

Manufacturer's name and address (if available)
Date of manufacture and/or lot number (if available)
Container size
Estimated amount in container (determine by visual inspection—do not remove lid if dry)
Concentration statements on label (if any)
Any evidence of container damage
Building name: _____
School address: _____
City or town: _____ Zip code: _____

trinitrophenol, picronitric acid, carbazotic acid, nitroxanthic acid, and phenoltrinitrate. Individuals conducting the inventory were cautioned to wear protective gloves, laboratory aprons, and eye protective equipment and to avoid unnecessary, or rough, handling of containers. Information requested of teachers included chemical quantity, container type and condition, physical state of chemical, and age if known.

By May 1979, approximately 200 of Iowa's 445 school districts and 6 of 20 community colleges reported the presence of various forms of picric acid. The individual quantities ranged from 1 ounce to 2 lb. The schools possessing the acid were widely dispersed across the state.

Since a great deal of ambiguous information was being circulated nationally concerning the properties and explosive potential of the material, more definitive guidelines had to be developed prior to asking teachers to assist in the redistribution or disposal operation.

By mid-May 1979, a 10-member, blue-ribbon committee representing state agencies who could provide applicable professional assistance was assembled by the State Science Consultant for the DPI in Des Moines. The agencies represented included: the State Fire Marshall's Office, Office of Disaster Services (ODS), Department of Environmental Quality (DEQ), University Hygienic Lab, Environmental Health Services—University of Iowa, Chemical Safety Committee—Iowa State University, Iowa Science Safety Task Force, Community Colleges, and Department of Public Instruction.

On conferring with chemical manufacturers, state chemists, and the EPA, guidelines were developed, to provide safe handling for local disposal of small quantities of the picric acid. Copies of the disposal instructions were developed on the basis of testing at the University of Iowa. Large quantities of the acid, excessively dry batches, or containers of the acid from which lids could not be removed (after 48 hours of soaking in water), were removed from schools individually by the state fire marshall's office, local sheriff's office, local fire department, or by the DEQ. In most cases, these containers of picric acid were remotely detonated in an environmentally approved landsite.

On May 10, 1979 and May 24, 1979, letters regarding picric acid stabilization and disposal were mailed to all secondary educational institution administrators, who had identified the acid in their science storerooms, providing directions both for short-term stability and ultimate disposal. Instructions were based on the fact that picric acid normally contains 10–20% water for stability. The content of these letters is presented in Sections 9.2.2 and 9.2.3.

As of this publication all identified picric acid has been eliminated from local schools, without incident, by means of the above procedure or by the DEQ or the local fire marshall's office.

9.2.2   Letter 1: Inventory of Picric Acid in Iowa Schools

Although picric acid is used predominantly in science classes, it *may* be utilized in art (staining glass) and crafts (tanning leather) classes also.

Picric acid normally contains 10–20% water for stabilizing. With the passage of time, such chemicals may become sufficiently dry to present potential explosive hazards. However, to date no such explosions have occurred without purposeful detonation.

In order to assess the quantities of picric acid presently in Iowa schools, and to initiate removal and/or dilution procedures the DPI is asking for your cooperation.

Things to Do: The following measures are recommended:

1. Avoid excessive movement or concussion of containers.
2. Carefully inventory your supplies (this is probably done regularly each year).

3. Complete the attached questionnaire and return (Table 9.1). Be sure to note *all* compounds and derivatives of picric acid [i.e., ammonium picrate (2,4,6-trinitrophenol, picro-nitric acid, carbazotic acid, nitroxanthic acid)].

4. Wear protective chemical gloves, laboratory apron, and eye protection as a precaution against skin contact.

5. Remove all oxidizable materials, finely divided metals, and alkaloids from the immediate vicinity of the picric acid.

6. Confirm that storage area is well ventilated and has restricted access.

If Spills Occur.   Absorb materials with sodium bicarbonate or sand–soda ash mixture (90:10 mixture). Store in covered glass container and call your local Department of Environmental Quality Spills Unit.

Things Not to Do.   The following measures must not be undertaken:

1. Do not open container if picric acid is dry (similar to dry salt or sugar). Tilt bottle. If crystals roll over each other, this may indicate a sufficiently dry state to be hazardous.

2. Do not attempt disposal in *any* manner.

3. Do not attempt dilution until adequate information is provided by chemical experts and manufacturers. (This will appear in future newsletters.)

9.2.3   Letter 2: Picric Acid Stabilization

Following conferences with representatives of the Department of Environmental Quality (DEQ), University of Iowa Hygienics Lab, Iowa State Fire Marshall's Office, University of Iowa Environmental Health Service, Office of Disaster Services, and the Iowa Science Safety Task Force, the enclosed guidelines were developed.

Picric acid normally contains 10–20% water for stabilizing. The following instructions are premised on increasing such stability while decreasing short-term potential hazards.

Things To Do: The following measures are recommended:

1. All preparations should be completed in a laboratory near a sink. Sand–soda ash mixture (10:1) should be available in case of accidental spill (refer to April 20 letter).

2. Wear rubber gloves, laboratory apron, and eye protection when handling the containers of picric acid.

3. All bottles containing the picric acid should be handled gently and cautiously.

4. Clean the external surface of the bottle by wiping with a damp cloth.

5. *Carefully* invert the bottle of picric acid in a beaker, or similar glass container, filled with tap water (Figure 9.1).

6. Cover beaker with glass plate to help assure immersion of the bottle and to help reduce water evaporation.

7. Clearly *label the beaker:* "PICRIC ACID—DO NOT DISTURB."

8. Inform custodial staff not to tamper with this setup.

9. Place the beaker on a shelf away from general access.

10. Do not be concerned with possible gas bubbles or water discoloration; these are both positive signs that water is entering the bottle and stabilizing the picric acid. Discolored water indicates that acid is entering the water. As this is a strong dye, do not allow contact with skin surface.

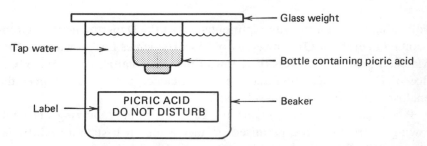

FIGURE 9.1   Submersion of a bottle of picric acid in water. The inverted position allows water to enter and remove sensitivity to explosion by moistening the chemical.

Spills. If a spill occurs *in the sink,* flush with copious amounts of water, or if there is a spill *on the counter or floor,* cover with sand–soda ash mixture; place glass cover over spill. In *both* cases, call the DEQ Spills Unit.

Things Not to Do. Do not remove lid from picric acid bottle, and do not attempt disposal. This procedure will help assure stability in addition to preparing the container for further dilution procedures.

9.2.4   Letter 3: Picric Acid Disposal for Iowa Schools

General Discussion. This disposal technique is essentially one of dilution followed by discharge to the sanitary sewer system. These instructions presuppose that the picric acid container has already been immersed in water in accordance with earlier instructions or that the acid is contained in a vessel that is easily opened.

Some consideration must be given to the concentration of picric acid solutions being discharged to the sanitary sewer. Picric acid is toxic to the biological organisms of a sewage treatment plant in concentrations greater than 200 ppm (200 mg/liter). It is safe to assume that at least a tenfold dilution will occur in the plumbing system of any metropolitan sewage treatment system. It is important, therefore, that the picric acid solution be less than 2000 ppm at the point of discharge (laboratory sink). The procedure described in the following paragraphs will ensure that.

If your school is not connected to a municipal sewage system but rather has its own septic system, do not follow this procedure. Instead, contact environmental authorities for specific instructions.

Useful Conversion factors. The units most commonly used in picric acid disposal are as follow:

$$1 \text{ pound} = 16 \text{ ounces} \qquad 1 \text{ ounce} = 24.4 \text{ g}$$
$$1 \text{ pound} = 454 \text{ g} \qquad 1 \text{ gallon} = 3.8 \text{ liters}$$

The solubility limit of picric acid in water at 25°C (77°F) is:

$$14,000 \text{ ppm} = 14,000 \text{ mg/liter} =$$
$$14 \text{ g/liter} = 53.2 \text{ g/gallon} =$$
$$0.12 \text{ pounds/liter} = 0.44 \text{ pounds/gallon}$$

Disposal. The following measures are recommended for picric acid disposal:

1. Picric acid is a strong mordant dye. Wear rubber gloves and laboratory apron. Wear eye protection. Picric acid is also toxic by inhalation (of dust), skin absorption, and ingestion. Wash any accidentally exposed skin immediately with soap and water. In case of accidental eye contact, flush immediately with water for at least 15 minutes *and* consult a physician.

2. Remove the picric acid container from the beaker of water in which it had been immersed (see Sec. 9.2.3). Rinse the exterior of the bottle in the sink with copious amounts of cold tap water.

3. Inspect the water in the immersion container. The intensity of yellow coloration is directly proportional to the picric acid concentration. If the immersion container is glass, proceed to step 4. If not, transfer the water into a suitable clean *glass* beaker (1 liter or larger) and then proceed to step 4.

4. Carefully open the picric acid container. *Note:* If the bottle will not open with hand force, even after immersion in water, *do not* apply excessive force by rapping the lid or using tools. Instead, contact the DEQ spills unit for further instructions.

5. Slowly add the picric acid to the immersion water with gentle stirring using a glass rod. Continue additions until the solubility limit is approached. *Do not* apply heat to enhance solubility. Follow the table below to determine *approximate* amounts of picric acid to reach solubility limits.

Container Volume	Weight of Picric Acid
1 liter	14 g
1 liter	0.03 lb
1 liter	0.5 ounces
1 gallon	53.2 g
1 gallon	0.12 lb
1 gallon	1.8 ounces

Note: All table values are approximate. Remember that you don't know the initial percent water of the solid picric acid (it may be as high as 15%). Remember also that immersion of the picric bottle in water has allowed entry of water into the bottle by capillary action, thus allowing hydration of the picric acid to some extent. It is not necessary, therefore, to weigh the picric acid prior to its addition to the water.

6. Open the *cold* water tap in the laboratory sink and verify that it is discharging at a rate of 10 liters/minute (2.6 gallons/minute) or more. *Note:* This discharge rate is assumed to be easily obtainable by most laboratory water taps. If your facility cannot attain this rate of discharge, adjust the following steps accordingly. *Do not* use hot water to obtain a flow increase.

7. Assuming your picric acid solution is at or near saturation, the acid concentration will be 10,000–14,000 ppm (at 25°C). With the cold water tap discharging at approximately 10 liters/minute, slowly and evenly discharge one liter of saturated picric acid solution into the sink over a 2-minute period. Under these conditions, the average picric acid concentration in the sink drain will be 476–666 ppm (10,000–14,000 mg of acid in 21 liters of water). Consider this to be a maximum discharge rate. Obviously, the slower the discharge rate of acid solution, the greater the dilution, the less probable the adverse impact on the sewage treatment plant. Feel free to lower the discharge rate of the acid and expand the dilution to the limits of practicality.

8. Continue by refilling the acid solution container with clean tap water and resuming additions of picric acid to that water until the solubility limit is again reached. Discharge this solution to the sink as in step 7. Repeat steps 7 and 8 until the entire supply of picric acid is depleted.

9. Wash all glassware and the sink with a mild bicarbonate solution, followed by a strong soap solution. Rinse with tap water, paying particular attention to removal of all splash residues on the laboratory counter and upper sink walls.

10. Dispose of the empty, rinsed, picric acid container in the trash. Glass-stoppered bottles can be returned to stock after thorough washing.

Finally, it is recommended that this disposal be done during peak flow to the sewage treatment plant, which in most cases is early morning (8:00–9:00 A.M.).

9.3 STATE-WIDE ASSESSMENT OF HAZARDOUS CHEMICALS

In August 1979 local school administrators and teachers again initiated numerous calls to the DPI and DEQ for assistance in disposal of other hazardous materials. Apparently, while conducting the picric acid inventory, teachers had discovered many other hazardous substances being warehoused in their chemical storerooms. On August 20, 1979, a letter concerning *Inventory and Disposal of Hazardous Chemicals* was sent by the State Science Consultant to all Iowa secondary educational institution administrators and science teachers asking them to inventory their storerooms for quantity, container type, and date of purchase of the chemicals in the following families: ethers; organic solvents; organic phosphates; organic acids; anhydrides; organic halogens; mineral acids and bases; inorganic hydrides and borohydrides; organic and inorganic peroxides; perchlorates; phosphorous compounds; cyano compounds; nitrates; azide and azocompounds; mercury compounds; alkaline earth metals; and alkali metals.

Approximately 400 schools identified 230 different chemicals of a potentially hazardous nature in their storerooms. From November 1979 through May 1980 the blue-ribbon state chemical safety committee reviewed and refined the hazardous chemicals list, weighing the academic value of the substance against its potential hazard. In addition to reviewing the National Fire Protection Agency (NFPA) guidelines (9), the following lists of hazardous substances were reviewed for application: Occupational Safety and Health Administration (OSHA) List of Defined Carcinogens (10); National Institute of Occupational Safety and Health (NIOSH) Summary of Recommendations for Health Standards (11); CHEM-13 Restricted Chemicals (12); National Science Teachers Association Table of Other Hazardous Substances (13); and the American Chemical Society List of Defined Teratogens (14). Material safety data sheets for specific chemicals were secured from chemical manufacturers and proved very helpful. Such data sheets provided information concerning the chemical's trade name, ingredients, physical data, fire and explosion hazard data, reactivity data, incompatibility, hazardous decomposition products, hazardous polymerization, potential spill or leak procedures, and disposal methods. Table 9.2 represents the synthesis of the Iowa inventory with these lists of hazardous substances. *It was concluded that these substances were more hazardous than their educational value could justify, and teachers should seriously consider removing them from their storerooms and from usage in their curriculum.*

9.3.1 The Problem Is Defined

Following delineation of a finite list of substances suggested for removal from schools, the problem became how to safely transport them from potentially 445 districts (785 secondary buildings) scattered thoroughout the state, to recycle and/or dispose of them, and to bear the cost of so doing. Chemical unknowns (unlabeled containers) were recognized as a separate problem and had to be addressed individually by the DEQ and/or fire marshall's office. This problem was compounded by new, more stringent handling, transportation, and disposal guidelines initiated by the EPA in November 1980 for certain quantities of certain substances (15).

9.3.2 Effects of New EPA Regulations

The Iowa DPI, in conjunction with the Iowa Council of Science Supervisors (CS^2), had recently completed a publication entitled *Better Science Through Safety*. (16). All secondary and community college science teachers were scheduled for in-service delivery of this publication between January and May 1981 at the 15 Area Education Agency offices located throughout the state. A large portion of the inservice addressed aspects of chemical safety. This seemed like a unique opportunity to increase teachers' awareness of science safety and to economically and legally dispose of the identified substances.

As the initial plan began to materialize, it became apparent that a diversity of state agencies would have to be coordinated. The governor's was the only office that had the authority to mobilize such agencies. On August 7, 1980, therefore, representatives from DPI and DEQ verbally outlined the plan below to the governor's administrative assistant. He suggested that the plan be submitted immediately in writing to the governor. On August 8, 1980 the following abridged plan was submitted to the office of Governor Robert D. Ray:

1. Teachers complete manifest of chemicals for disposal—forward to DEQ.

2. Packaging instructions (pertinent to manifest) and Department of Transportation packaging labels forwarded to teachers; pack-

TABLE 9.2 Chemicals Suggested for Removal (Dependent on Age and Quantity) from Iowa Schools

2,4-Dichlorophenoxy acetic acid (2,4-D)
2-Acetylaminofluorene
Acrylonitrile
α-Naphthylamine
Ammonium perchlorate
4-Aminobiphenyl
Anhydron (cyclothiazide)
Arsenic (inorganic)
Arsenic acid
Arsenic sulfide
Arsenic trioxide
Arsenic pentoxide
Arsenic trichloride
Arsenious acid
Asbestos (friable)
Barium chromate
Benzene
Benzidine (all derivatives)
Benzoyl peroxide
Benzyl chloride (chlorotoluene)
Beryllium
Beryllium nitrate
β-Naphthylamine
β-Propiolactine
Biphenyls (all derivatives)
Bischloromethylether
Bromotoluene
Calcium carbide
Carbamates (all derivatives)
Chloral hydrate
Chlorosulfonic acid (sulfuric chlorohydrin)
1-Chloro-2,4-dinitrobenzene
25WP (wettable powder) Cythion (Malathion)

1,2-Dibromo-3-chloropropane (DBCP) dibromochloropro-pane
3,3-Dichlorobenzidine (all salts)
1,2-Dichloroethyl ether
DDT
Dieldrin
Diethyl sulfate
Dimethyl sulfate
4-Dimethylaminoazo-benzene
Ethyleneimine
Ethylene chlorohydrin
Ethylene dibromide (EDB) (1,2-dibromoethane)
1-Fluoro-2,4-dinitrobenzene
Formaldehyde (not formalin)
Hydrazine
Hydrofluoric acid
Hydrogen cyanide (hydrocyanic acid)
Isobutyl mercaptan (2-methyl-1-propanethiol)
Lead arsenate
Lithium aluminum hydride
Malathion
4,4-methylene bis (2-chloroaniline)
Mercury alkyls
Mercury cyanide
Methylchloromethyl ether
Monochloroacetic acid (chloroacetic acid)
Nicotine (nicotine sulfate)

Nitrilotriacetic acid
4-Nitrobiphenyl (ONB)
Nitrocellulose
Nitrogen triiodide
N-Nitroso-dimethylamine
Nitrosophenols (meta- and para-phenols)
Nitrotoluene (all isomers)
Perchloric acid
Perosmic acid
Phenanthrene
Phosphides (all metal)
Phosphorous (white, yellow)
Phosphorous pentoxide (phosphorus anhydride)
Picric acid
Potassium
Potassium amide
Potassium azide
Potassium cyanide
Silver cyanide
Sodium arsenite
Sodium azide
Thallium (all compounds)
Tetraethyldithiopyro-phosphate (TEDP)
Thermite (igniting mixture—aluminum filings and iron oxide)
Thionyl chloride
Vinyl chloride monomer (chloroethene)
Zinc chromate

210

aging guidelines designed to help avoid packing incompatible substances in the same or proximate packages.

3. Teachers transport chemicals to science safety inservice at Area Education Agency office on date specified.

4. Iowa National Guard trucks accept manifest chemicals and transport to single collection center, University of Iowa.

5. Commercial disposal company accepts chemicals at the University of Iowa and transports to Illinois for disposal and/or recycling.

By October 1980, we were informed that the governor's office could not mobilize the Iowa National Guard unless a state of emergency existed. That did not, however, prevent local guard units from transporting such substances as a training exercise or a community service. Severe state economic constraints also made funding of such a program impossible. Furthermore, union contracts and vehicle insurance restrictions prevented other education-related vehicles from hauling such substances across state highways.

It also became apparent that the new EPA regulations concerning generation, handling, transport, and disposal of hazardous substances would be in effect (November 1980) before a new plan could be formulated and implemented. The DEQ and DPI worked closely with the EPA from September 1980 to January 1981. It finally became apparent that standards regulating the handling, transportation, and disposal of certain hazardous substances in Section 261.33 of the *Federal Register* (17) (15 of which did appear on the inventories of Iowa schools) could not be suspended for a one-time cleanup for Iowa's educational institutions. This decision, in conjunction with increasing financial restrictions and transportation problems, resulted in the suspension of the original plan and the development of an alternative.

9.4 AN ALTERNATIVE DISPOSAL PLAN FOR IOWA SCHOOLS

This alternative plan was premised on the assumption that many of the hazardous substances identified in Iowa schools could not be safely and equitably transported according to EPA regulations. It was decided that the next best alternative would be to provide the consultive expertise of specially trained university chemists to teachers through a telephone network.

Chemists from each of the Iowa Regent's institutions (state universities) and state hygenic labs were selected to attend a 1-day training program at the University of Iowa, Department of Environmental Health. At that meeting the hazardous chemicals inventories from Iowa's schools were reviewed and common disposal instructions were synthesized for those substances that could be safely and legally disposed of by science teachers. It was suggested by teachers that any disposal techniques should be reliable, safe, inexpensive, and attainable with the equipment limitations of the typical rural Iowa school. The techniques finally synthesized from a diversity of reliable resources were therefore limited to evaporation, dilution, neutralization, open burning, or redistribution to local industries or colleges. Certain acutely hazardous materials (EPA 261.33) that presented unusual disposal problems were referred to the Iowa DEQ or Iowa State Fire Marshall for disposal or redistribution.

A telephone response form (Table 9.3) was developed to assure that the chemistry teachers attempting disposal, according to university chemist instructions, were professionally qualified to do so; that they had the proper equipment to safely conduct the procedures; that they were not exceeding safe limits for disposal of chemicals at sewage treatment plants or for septic tank systems; and that the university chemists knew certain characteristics of the chemicals proposed for disposal such as age, quantity, container type, and physical state.

A telephone response was kept for each responding school. Duplicate copies of the forms were forwarded to the DEQ for each school that had materials for which the DEQ would have to arrange disposal.

The Iowa Regent's institution chemists were selected on the basis of expertise, willingness to assist in this effort, and their liability protection under the state system.

In addition to training from the Environmental Health Department of the University of Iowa, the chemists received applicable EPA standards information from DEQ chemists.

On April 20, 1981 memos entitled "Elimination of Unwanted Chemicals from Science Storerooms" were sent from the DPI to all Iowa school administrators and to all AEA science consultants outlining the above plan and the contacts for each state area (Figure 9.2). In addition, announce-

TABLE 9.3   Telephone Response Form[a]

Date
Contact person
School and address
Telephone number
Professional background of contact person
   (educational degree, chemistry training, etc.)
School facilities, equipment
  Functional exhaust hood
  Sewer system
    Septic tank
    Proximity to sewage treatment plant
  Goggles
  Gloves
  Apron
  Face shield
Chemical data

	Name of Substance	Physical State	Age	Container Type, Condition	Quantity	Proposed Action
1						
2						
3						
4						
5						
6						
7						
8						

[a]This information, gathered by consulting professional chemists, was used to identify unwanted chemicals for waste disposal recommendations.

ments of the plan were placed in the May issue of *Science Spectrum* (monthly publication of the DPI science consultant to all science teachers), the *DPI Dispatch* (quarterly newsletter to all Iowa educators), several newspapers, and numerous radio and television stations. The next two sections contain the content from actual memos and inventory sheets used in the "Iowa Experience."

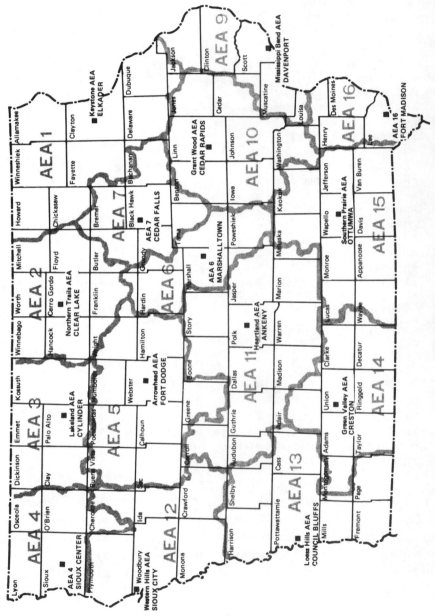

FIGURE 9.2

9.4.1. Memo 1: Inventory and Disposal of Hazardous Chemicals

In order to provide schools with information for dilution and disposal of chemicals, it is necessary to know what those chemicals are, as well as their physical state, quantity, and age. In addition, we would like to caution teachers against storage or retention of unlabeled or incompletely labeled materials. Whenever possible, retain manufacturers storage containers. If original containers are damaged, cautiously transfer to appropriate alternative container and label completely. Do not purchase more than needed for 1 year.

The enclosed inventory (Table 9.4) was developed to provide nonprescriptive guidelines for assessing chemical stores of suspect chemicals, compressed gases, and containers of unknown or unlabeled substances.

Please reproduce the inventory for *each* secondary (grades 7–14) building in your district and have all storerooms checked for these hazardous chemicals. On compilation of all data, specific instructions concerning dilution and/or appropriate disposal techniques will be forwarded to you.

Should you, or your staff, have future *technical* questions concerning chemicals, please contact the regional Department of Environmental Quality office (see map given in Figure 9.3) in your area. You are also encouraged to contact the office serving your county to set up appointments to review the reference materials, you may decide that one or more of the volumes is appropriate for your teachers' reference library.

←

FIGURE 9.2 Map of Area Education Agencies and contact institutions for assistance with hazardous chemicals.

Contact Institution	Telephone[a]	Area Education Agency
University of Northern Iowa		1, 2, 6, 7
Iowa State University		3, 4, 5, 11, 12, 14
University Hygiene Laboratory (Des Moines)		3, 4, 5, 11, 12, 14
Department of Environmental Quality		3, 4, 5, 11, 12, 14
University of Iowa		9, 10, 15, 16
University Hygiene Laboratory (Iowa City)		9, 10, 15, 16

[a]Actual telephone numbers of Contact Institutions were supplied to each Iowa school but omitted here out of courtesy.

TABLE 9.4 Hazardous Chemical Inventory Sheet Used in Iowa Schools

Hazardous Chemical Inventory

District Name _____

School Name _____

Instructions

1. All storerooms should be checked for these hazardous chemicals.
2. ____ Check if you do not have any of these hazardous chemicals on hand.
3. Make a copy of the completed form and retain for future reference relating to disposal of these chemicals.
4. Return the completed form to your superintendent.
5. *Do not dispose of any of these hazardous chemicals until you have been contacted.*

Chemical Family	Names of Chemicals on Hand	Quantity (Ounces or Grams)	Container (Box, Bottle, Can, etc.)	Physical State of Chemical (Solid, Liquid, Gas)	Date of Purchase
I. Alkaline earth metals (zero valent); e.g., sodium, potassium)	1. ____ 2. ____ 3. ____ 4. ____ 5. ____	____ ____ ____ ____ ____	____ ____ ____ ____ ____	____ ____ ____ ____ ____	____ ____ ____ ____ ____
II. Organic halogens also have applications as pesticides (e.g., hexaclorobenzene, sodium fluoracetate)	1. ____ 2. ____ 3. ____ 4. ____ 5. ____	____ ____ ____ ____ ____	____ ____ ____ ____ ____	____ ____ ____ ____ ____	____ ____ ____ ____ ____

216

III. Organic phosphates (thio-phosphates; e.g., mala-thion, parathion)

1. — — — — — — — — — —
2. — — — — — — — — — —
3. — — — — — — — — — —
4. — — — — — — — — — —
5. — — — — — — — — — —

IV. Azides and azo com-pounds (including hydra-zoic acid; e.g., azide, azobenzene)

1. — — — — — — — — — —
2. — — — — — — — — — —
3. — — — — — — — — — —
4. — — — — — — — — — —
5. — — — — — — — — — —

V. Percholorates (including perchloric acid; e.g., so-dium perchlorate)

1. — — — — — — — — — —
2. — — — — — — — — — —
3. — — — — — — — — — —
4. — — — — — — — — — —
5. — — — — — — — — — —

VI. Organic and inorganic peroxides (including hydrogen peroxide over 70%; e.g., benzoyl peroxide)

1. — — — — — — — — — —
2. — — — — — — — — — —
3. — — — — — — — — — —
4. — — — — — — — — — —
5. — — — — — — — — — —

VII. Cyanides and nitriles (e.g., potassium cyanide, acrylonitrile)

1. — — — — — — — — — —
2. — — — — — — — — — —
3. — — — — — — — — — —
4. — — — — — — — — — —
5. — — — — — — — — — —

217

TABLE 9.4 (Continued)

Chemical Family	Names of Chemicals on Hand	Quantity (Ounces or Grams)	Container (Box, Bottle, Can, etc.)	Physical State of Chemical (Solid, Liquid, Gas)	Date of Purchase
VIII. Ethers (e.g., diethylether)	1. ——— 2. ——— 3. ——— 4. ——— 5. ———	——— ——— ——— ——— ———	——— ——— ——— ——— ———	——— ——— ——— ——— ———	——— ——— ——— ——— ———
IX. Organic and inorganic hydrides (including hydrazine; e.g., lithium aluminum hydride, silane)	1. ——— 2. ——— 3. ——— 4. ——— 5. ———	——— ——— ——— ——— ———	——— ——— ——— ——— ———	——— ——— ——— ——— ———	——— ——— ——— ——— ———
X. Mercury and its compounds (e.g., mercuric chloride)	1. ——— 2. ——— 3. ——— 4. ——— 5. ———	——— ——— ——— ——— ———	——— ——— ——— ——— ———	——— ——— ——— ——— ———	——— ——— ——— ——— ———
XI. Beryllium and its compounds (e.g., beryllium hydroxide)	1. ——— 2. ——— 3. ——— 4. ——— 5. ———	——— ——— ——— ——— ———	——— ——— ——— ——— ———	——— ——— ——— ——— ———	——— ——— ——— ——— ———

XII. Phosphorus (white, yellow)

1. _____
2. _____
3. _____
4. _____
5. _____

XIII. Arsenic and its compounds (e.g., arsenic trichloride, lead arsenate)

1. _____
2. _____
3. _____
4. _____
5. _____

	Container Size[a]	Estimated Quantity in Container (Full, Half-Full, etc.)	Container (Box, Bottle, Can, etc.)	Physical State of Chemical (Solid, Liquid, Gas)	Comments
XIV-1	_____	_____	_____	_____	_____
XIV-2	_____	_____	_____	_____	_____
XIV-3	_____	_____	_____	_____	_____
XIV-4	_____	_____	_____	_____	_____
XIV-5	_____	_____	_____	_____	_____
XIV-6	_____	_____	_____	_____	_____
XIV-7	_____	_____	_____	_____	_____
XIV-8	_____	_____	_____	_____	_____
XIV-9	_____	_____	_____	_____	_____

[a]Please take time to identify unlabeled containers. Place a piece of masking tape on the container and number them sequentially so we may refer to them at a later date.

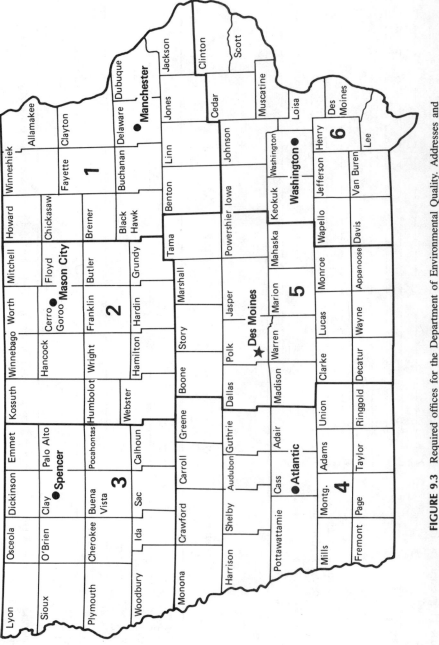

FIGURE 9.3 Required offices for the Department of Environmental Quality. Addresses and telephone numbers supplied to each Iowa school are omitted here out of courtesy.

220

9.4.2 Memo 2: Elimination of Unwanted Chemicals from Science Storerooms

Potentially hazardous chemicals are by-products of many processes characteristic of a highly technological society. Science curricula in secondary schools of such a society also periodically involve the mixture and/or production of some sophisticated chemical substances.

Following three inventories (1978, 1980, and 1981) of Iowa secondary schools conducted by the Department of Public Instruction, it was discovered that there are indeed many potentially hazardous chemicals in science storerooms. In many instances these chemicals had accumulated over many years of science instruction. Teachers should be encouraged to conduct regular inventories of their chemical storerooms and remove unnecessary substances. Especially volatile or hazardous chemicals should be ordered only in sufficient quantity to be utilized with one academic year. If either sufficient safety equipment or teacher knowledge of all chemical properties of the substances are lacking, serious consideration should be given to removing them from the science program.

New EPA regulations concerning chemical manifests as well as high transportation costs make state collection and disposal of all hazardous substances in Iowa schools impossible. In an effort to assist schools in handling, storage, and disposal of these chemicals a state science safety committee was developed, with representation from the DPI, DEQ, Iowa State Fire Marshall's Office, University Hygienic Laboratory, Environmental Health Department of the University of Iowa, and the Department of Transportation (DOT). Following careful review of the Iowa inventories it was decided that, since schools had such a diversity of chemicals it would also be impossible to address all needs through a written communiqué. Chemists at the regents institutions, University Hygienic Laboratory, and DEQ thus received special training in handling and disposal of hazardous chemicals in order to answer specific questions which science teachers might have. (The cadre of chemists is outlined in Figure 9.2)

Procedures. The following steps are recommended:

1. Each school submits a list (in writing or by telephone) of unwanted chemicals to one of the professional chemists identified. These chemists will return suggestions to you for neutralizing,

oxidizing, reducing, and so on, those chemicals that you can safely handle and that can enter the environment safely. The list of chemicals remaining after this exchange will be forwarded to the Department of Environmental Quality (DEQ) by the professional chemists.

2. The DEQ will act on the list of chemicals according to the following scheme:

    a. All shock-sensitive, flammable, and potentially explosive materials will be referred by DEQ to the Iowa State Fire Marshall's Office for disposal. The school will segregate these chemicals into containers, as instructed, and pack them to prevent breakage during transportation. A field agent will come directly to the school and transport the substances to an approved, remote area for detonation and ignition.

    b. The list of remaining materials will be processed for acceptance at a local landfill. A Special Waste Authorization (SWA) will be issued by DEQ to each school for disposing of the chemicals with specific handling instructions. It is the responsibility of each school to get these materials to the landfill.

    c. Acutely hazardous substances, as defined by the rules of the Resource Conservation and Recovery Act (RCRA), may be identified as "not controllable" in Iowa; each school with these substances may be notified by DEQ to contact chemical suppliers, or university laboratories, or hazardous waste repositories in another state for assistance in disposing of these substances. Each school would be responsible for this contact.

Each of these steps is necessary to protect the health of individuals who handle the materials as well as to prevent indiscriminant introduction of potentially harmful substances into the environment.

9.5   RESULTS OF THE PROGRAM

The response of school administrators, teachers, and the media to this plan was both immediate and positive. The only real criticism was that it seemed to take so long to arrive at such a simple and pragmatic solution.

Between May 1981 and July 1981 approximately 100 of Iowa's 445 school districts requested assistance in disposal of chemicals through this system.

Approximately 20 of the school districts possessed chemicals of an acutely hazardous nature and were referred to the DEQ for assistance. All schools appeared very satisfied with the consultive assistance they received. Considering that most Iowa schools terminate their regular academic year in early June, that few schools may have had any of the hazardous materials, that some schools may have undertaken their own disposal operations without consultive assistance, and that this was a difficult time of year to address such problems, a 22% response was considered favorable.

In September the plan was again communicated to all Iowa science educators and administrators by a mechanism similar to those discussed above. By December 1981, nearly 30% of Iowa's science teachers had safely purged a diversity of hazardous chemicals from their storerooms. Those substances categorized as acutely hazardous by the university chemists were referred to the DEQ and fire marshall's office and disposed of in a safe and approved manner. The state fire marshall has six regional marshalls scattered throughout the state. To facilitate safe transport of these acutely hazardous chemicals from schools to disposal sites by regional fire marshalls, DEQ chemists reviewed each school's inventory and divided them into compatible lots. Packaging instructions were also provided to the fire marshalls.

9.6 SUMMARY AND CONCLUSIONS

This plan has proved successful for one state. It was not designed to be applied, without modification, in other states. The assessment instruments and basic format (summarized below), however, may be modified to meet the needs of a diversity of states with similar concerns and limited budgets. It may also make it possible for other states to anticipate, and avoid, some of the problems encountered in Iowa.

1. Identify a state coordinating agency. The state department of education represents a nonthreatening agency that already has established contacts with schools. Involve other state agencies as necessary.

2. Assess the risk/benefit value of the types of chemicals currently in use in schools. State chemists, chemical manufacturers, and emergency services agencies should be involved.

3. Assess the magnitude of the problem. Inventory all educational institutions, collecting such information as the chemical's age, the manufacturer's lot number, the quantity, the physical state, and the container condition.

4. Contact chemical manufacturers of the hazardous chemicals and request their material safety data sheets for specified chemicals. Request the manufacturer's assistance in handling and disposal of the substances.

5. Develop listings of hazardous chemicals suggested for renewal from schools and safe mechanisms for their redistribution or disposal. Federal EPA and state laws and guidelines should be reviewed for applicability.

6. Develop a cadre of chemical experts who can be instructed to provide established technical assistance to teachers. Each chemist should be assigned a specific state region of responsibility.

7. Communicate the chemical redistribution and/or disposal plan, list of suspect chemicals, and names of chemists who can provide technical assistance to science teachers throughout the state. Utilize a diversity of communications mechanisms to assure redundancy.

8. Develop instruments for use by chemists in providing technical assistance, via telephone, to teachers. Include name of contact person; school; telephone number; teacher's academic preparation in chemistry; proximity of school to sewage treatment plant or septic tank; safety equipment in school laboratory; chemical age, physical state, quantity, container condition; and proposed action.

9. Communicate names of acutely hazardous chemicals to state environmental protection agency, or fire marshall, so that they may arrange for pickup and special disposal in an environmentally approved manner. Chemists should prepare packaging instructions for each school's acutely hazardous chemicals that will assure safe transportation to disposal sites.

10. Remind schools of this assistance program at the beginning of each academic semester. The service should be maintained for 2–5 years to help assure safe elimination of hazardous chemicals and to remind teachers of the necessity of conducting regular inventories of their chemical storerooms.

11. Arrange for safety in-service programs for science teachers.

There is little doubt that addressing problems affiliated with the safe generation, transportation, recycling, and/or disposal of hazardous substances will be one of this technological society's concerns in the immediate and long-term future. The situation in secondary schools reflects only a microcosm of this dynamic problem. The plan outlined in this chapter for Iowa schools has proven so successful that it will remain in effect for the next 5 years in the event that other chemical problems and/or questions arise.

If individuals, institutions, and states are to successfully address such issues, it will be necessary to develop an entirely new attitude toward our affluent nature, our view of our finite resources, and our concerns for the children in local and cosmopolitan communities.

For educational institutions specifically, scientists and educators must become more knowledgeable and critical of the materials utilized in the science curriculum. In addition, more stringent guidelines must be established by science educators concerning purchasing philosophy, inventory control, and regular disposal of unwanted chemical substances. Inventories should also be checked regularly to avoid repetition of a similar situation in the future. Cooperative efforts by appropriate state agencies can address many of these concerns and assist science educators in averting future tragedies in educational settings.

REFERENCES

1. C. R. Coble and P. B. Hounsell, "A Framework for Evaluating Chemical Hazards," *The Science Teacher*, **47**, 36–40 (1980).

2. R. D. Macomber, "Chemistry Accidents in High School," *J. Chem. Ed.*, **38**, 367–368 (1961).

3. M. Reutter, "Rand McNally to Pay Damages in School Textbook Misshap," *Publishers Weekly* (September 26, 1980).

4. *Mastrangelo vs West Side Union High School*, 42 Pacific Reporter, 2d Series, pp 634 (1975).

5. Memo from Ken Schaefer (Science Supervisor, Mason City Schools) to Jack A. Gerlovich (Science Consultant, Iowa Department of Public Instruction), October 5, 1979.

6. J. J. McDermott and I. T. Edgar, "The Picric Acid Episode—an 'Isolated Incident' That Wasn't," *The Science Teacher*, **46**, 35–36 (1979).

7. G. W. Stacey, ACS Tenth Biennial Education Conference: "Safety and Health in the Academic Laboratory: Recommendations by the Participants," *J. Chem. Ed.*, **52**, 91–93 (1979).

8. R. D. Benton, "Picric Acid Means Danger in School Lab," Department of Public Instruction—*OUTREACH*, **6**, (II), 6, (April 20, 1979).

9. Fire Protection Guide on Hazardous Materials, National Fire Protection Association, Boston, Massachusetts.

10. *NIOSH/OSHA Pocket Guide to Chemical Hazards*, NIOSH Publication No. 78–210. 1979.

11. "Summary of NIOSH Recommendations for Occupational Health Standards," October 1978; *CHEM-13 News,* **110**, 9–11 (January 1980).

12. "Restricted Chemicals," *CHEM-13 News,* **117**, 3–9 (November 1980).

13. P. B. Hounsell, "A Framework for Evaluating Chemical Hazards," *The Science Teacher*, **47**, 36–40 (1980).

14. Committee on Chemical Safety, *Safety in Academic Chemistry Laboratories*, 3 ed., American Chemical Society, Washington, DC, 1976.

15. *Federal Register*, Environmental Protection Agency; *Hazardous Waste Management System: Identification and Listing of Hazardous Waste*, Vol. 45, No. 229, pp. 178524–78550, November 25, 1980.

16. J. A. Gerlovich, G. E. Downs, F. W. Starr, et al., *Better Science Through Safety*, Iowa State University Press, 1981.

17. *Federal Register,* Environmental Protection Agency; *Hazardous Waste and Consolidated Permit Regulations,* Vol. 45, No. 98, pp. 33125–33127, May 19, 1980.

18. State of Iowa, *School Laws of Iowa*, compiled from the Code of Iowa and Acts of the General Assembly, 1980.

19. *Science Spectrum*, Iowa Department of Public Instruction, Des Moines, Iowa, May 1981.

20. W. Smith, "Chemists to Help Science Teachers," *DISPATCH*, Iowa Department of Public Instruction, Des Moines, Iowa, April–May 1981, Vol. 10, No. 7p. 8.

APPENDIXES

APPENDIX 1

J. T. BAKER SAF-T-DATA LABELING SYSTEM

The J. T. Baker Chemical Company has pioneered a unique labeling system that provides an excellent source of safety information for each of the over 3000 laboratory chemicals produced. The label has been reviewed and endorsed by representatives of both the Occupational Safety and Health Administration (OSHA) and the National Institute of Occupational Safety and Health (NIOSH). Designed to serve as a hazard communication for the user of the chemical, the label provides several pieces of information:

1. *Storage Color Coding.* The upper left-hand corner of the label is color coded to recommend proper storage of the material with other and similar products. Below is a list of colors and the storage coding:

Blue: Health hazard. Store in a secure poison area.

Yellow: Reactivity hazard. Store separately and away from flammable
           or combustible materials.

Red: Flammable hazard. Store in a flammable liquid storage area.

White: Contact hazard. Store in a corrosionproof area.

Orange: Substances with no rating higher than 2 in any hazard cate-
           gory. Store in a general chemical storage area.

2. *Numerical Hazard Code.* Substances are rated on a scale of 0 (nonhazardous) to 4 (hazardous) in each of the four hazard categories (see Figure A1.1).

3. *Universal Pictograms for Laboratory Safety Equipment.* The pictograms suggest proper personal protective equipment and clothing to be used by anyone handling the substance (see Figure A1.2).

4. *Hazard Alert Information.* These data include DOT and IMO classifications, the CAS number, availability of Material Safety Data Sheets, EPA hazardous waste code, recommended spill cleanup materials, and the NFPA diamond with ratings from NFPA 704.

Figures A1.3 and A1.4 show two examples of the J. T. Baker SAF-T-DATA chemical label for hydrochloric acid and sodium hydroxide. Since

Since the time of this writing, two other chemical companies have implemented similar safety labelling on their brand of laboratory chemicals; Fisher Scientific, a division of Allied Chemical Corporation, advertises its **ChemAlert**™ *system, while Mallinckrodt, Inc. has more recently introduced its* **LabGuard**™ *system.*

NUMERICAL HAZARD CODE

Substances are rated on a scale of 0 (non-hazardous)
to 4 (extremely hazardous) in each of four hazard categories:

- **Health hazard**   - the danger or toxic effect a substance presents if inhaled, ingested, or absorbed.
- **Flammable hazard** - the tendency of the substance to burn.
- **Reactivity hazard** - the potential of a substance to explode or react violently with air, water or other substances.
- **Contact hazard**   - the danger a substance presents when exposed to skin, eyes, and mucous membranes.

Rating Scale

4	3	2	1	0
Extreme	Severe	Moderate	Slight	None*

*No scientific data in the standard references that suggests the substance is hazardous.

HAZARD SYMBOL

A substance rated 3 or 4 in any hazard category will also display a hazard symbol. These easy-to-understand pictograms emphasize the serious hazards related to a substance:

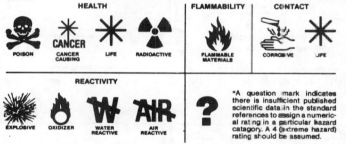

*A question mark indicates there is insufficient published scientific data in the standard references to assign a numerical rating in a particular hazard category. A 4 (extreme hazard) rating should be assumed.

FIGURE A1.1 Explanation key for the numerical hazard code for J. T. Baker SAF-T-DATA™ labeling systems.

LABORATORY PROTECTIVE EQUIPMENT

This series of pictograms suggests the personal protective clothing and equipment recommended for use when handling the substance in a <u>laboratory situation</u>. The pictograms relate to the combination of hazards presented by the substance.

The stop sign indicates the substance represents a special extreme hazard and the MSDS and other references should be consulted before handling.

FIGURE A1.2 Explanation key for laboratory protective equipment for J. T. Baker SAF-T-DATA™ labeling system.

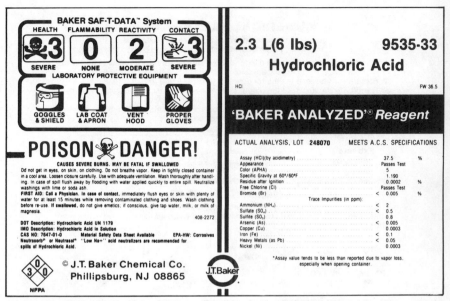

BAKER SAF-T-DATA™ System

HEALTH	FLAMMABILITY	REACTIVITY	CONTACT
3	0	2	3
SEVERE	NONE	MODERATE	SEVERE

LABORATORY PROTECTIVE EQUIPMENT

GOGGLES & SHIELD | LAB COAT & APRON | VENT HOOD | PROPER GLOVES

POISON DANGER!

CAUSES SEVERE BURNS. MAY BE FATAL IF SWALLOWED

Do not get in eyes, on skin, on clothing. Do not breathe vapor. Keep in tightly closed container in a cool area. Loosen closure carefully. Use with adequate ventilation. Wash thoroughly after handling. In case of spill flush away by flooding with water applied quickly to entire spill. Neutralize washings with lime or soda ash.
FIRST AID: Call a Physician. In case of contact, immediately flush eyes or skin with plenty of water for at least 15 minutes while removing contaminated clothing and shoes. Wash clothing before re-use. **If swallowed,** do not give emetics. If conscious, give tap water, milk, or milk of magnesia.

408-2272

DOT Description: Hydrochloric Acid UN 1179
IMO Description: Hydrochloric Acid in Solution
CAS NO: 7647-01-0   Material Safety Data Sheet Available   EPA-HW: Corrosives
Neutrasorb® or Neutrasol® ''Low Na+'' acid neutralizers are recommended for spills of Hydrochloric Acid.

© J.T. Baker Chemical Co.
Phillipsburg, NJ 08865

J.T.Baker

NFPA

2.3 L(6 lbs)   9535-33
Hydrochloric Acid

HCl   FW 36.5

'BAKER ANALYZED'® Reagent

ACTUAL ANALYSIS, LOT **248070**   MEETS A.C.S. SPECIFICATIONS

Assay (HCl)(by acidimetry)	37.5	%
Appearance	Passes Test	
Color (APHA)	5	
Specific Gravity at 60°/60°F	1.190	
Residue after Ignition	0.0002	%
Free Chlorine (Cl)	Passes Test	
Bromide (Br)	< 0.005	%

Trace Impurities (in ppm):

Ammonium (NH₄)	< 2
Sulfate (SO₄)	< 0.5
Sulfite (SO₃)	0.8
Arsenic (As)	< 0.005
Copper (Cu)	0.0003
Iron (Fe)	< 0.1
Heavy Metals (as Pb)	< 0.05
Nickel (Ni)	0.0003

*Assay value tends to be less than reported due to vapor loss, especially when opening container.

FIGURE A1.3 A representative SAF-T-DATA™ label for hydrochloric acid. (Courtesy of J. T. Baker Chemical Company.)

BAKER SAF-T-DATA™ System

HEALTH	FLAMMABILITY	REACTIVITY	CONTACT
3	0	2	4
SEVERE	NONE	MODERATE	EXTREME

LABORATORY PROTECTIVE EQUIPMENT

GOGGLES & SHIELD | LAB COAT & APRON | VENT HOOD | PROPER GLOVES

POISON DANGER!

CAUSES SEVERE BURNS. MAY BE FATAL IF SWALLOWED.

Do not get in eyes, on skin, on clothing. Avoid breathing vapor or mist. Keep in tightly closed container. In case of spill flood with water.
FIRST AID: Call a Physician. In case of contact, immediately flush eyes or skin with plenty of water for at least 15 minutes while removing contaminated clothing and shoes. Wash clothing before re-use. **If swallowed, if conscious,** give water with large amounts of diluted vinegar, lemon or orange juice. Follow with milk or whites of eggs beaten with water. Apply artificial respiration if not breathing. Keep patient warm and quiet.
DO NOT STORE BELOW 12°C(54°F)

408A-2338

DOT Description: Sodium Hydroxide, Solution UN 1824
IMO Description: Sodium Hydroxide, Solution
CAS NO: 1310-73-2   Material Safety Data Sheet Available   EPA-HW: Corrosives
Neutracit-2® Caustic Neutralizer is recommended for spills of Sodium Hydroxide.

© J.T. Baker Chemical Co.
Phillipsburg, NJ 08865

J.T.Baker

4 L   3727-3
Sodium Hydroxide
50% Solution (w/w)

NaOH   FW 40.0

'BAKER ANALYZED'® Reagent

ACTUAL ANALYSIS, LOT **249816**

Assay (NaOH)(by acidimetry)	50	%
Sodium Carbonate (Na₂CO₃)	0.06	%
Chloride (Cl)	< 0.001	%
Sulfate (SO₄)	0.001	%
Ammonium Hydroxide Precipitate	0.006	%
Heavy Metals (as Ag)	0.0005	%
Potassium (K)(by FES)	0.002	%

Trace Impurities (in ppm):

Nitrogen Compounds (as N)	2.5
Phosphate (PO₄)	< 1.2
Iron (Fe)	2.5
Nickel (Ni)	< 2.5

FOR DISPLAY ONLY

FIGURE A1.4 A representative SAF-T-DATA™ label for 50% sodium hydroxide. (Courtesy of J. T. Baker Chemical Company.)

both of these chemicals present a contact hazard, the storage color coding in the upper left-hand corner is white. Both of these chemicals should be stored in a corrosionproof area. Note the diagonal bars in the storage rectangle on the "Sodium Hydroxide" label in Figure A1.4. These bars on this label are designed to represent an incompatibility with other chemicals with a *blank* white storage color coding. Thus "Hydrochloric Acid," with a blank white storage coding, and "Sodium Hydroxide," with a diagonal bar white storage coding are indicated as incompatible and should be kept away from each other in the corrosionproof area. The "blank" and "diagonal bar" method is used to indicate incompatibilities in each of the four hazard storage classes. Another example of this distinction is oxidizer and reducing agents in the Reactivity Hazard storage area.

The J. T. Baker SAF-T-DATA labeling system represents the first chemical supplier's commitment to transmitting organized hazard information directly on the label. Such a commitment can only help to improve the safe operation of a chemical storeroom.

APPENDIX 2

PLANNING FOR PURCHASING FOR CHEMICAL STORAGE

JOHN BEQUETTE
University of Missouri, Columbia, Missouri

Each year millions of dollars are spent to purchase chemicals for research, education, and manufacturing. Of these, many dollars' worth of chemicals are wasted due to poor planning and improper storage and lack of safety considerations.

The keys to minimizing the amount of wasted dollars are: good planning based on accurate information such as previous usage rates, projected future usage rates, available storage space for different types of chemicals, safety regulations concerning storage of hazardous chemicals, ability of personnel to handle certain size containers of different chemicals, the economic feasibility of purchasing in large quantities, and the method and time required to replenish in-house stocks.

Planning should begin with management or administration defining clear-cut areas of responsibilities and purchasing procedures. Strict enforcement in these areas will eliminate delays in ordering and deliveries, duplicate shipments, wrong addresses on shipments, and confusion in accounting areas. All of these can be equated into lost dollars and cents by the work-hours and expense incurred to correct errors.

Each of the areas needed for proper planning, as listed above, must be considered alone and in conjunction with all other items, starting with previous usage rates obtained from records of past purchases and consumption rates. Accurate records are a necessity for accurate planning. (Computer systems covered in Chapters 5 and 8 can be an excellent tool in the storage of this information, which can be made readily available to management and purchasing.)

Projected usage may be obtained from investigators or from increases in number of staff using chemicals or common solvents. Changes in procedures in industry or research or in projected enrollment can also be used to project needs.

Available storage space is a very important item in the planning for purchasing as the shifting or relocating of items from normal storage areas may create a burden on buildings as well as personnel. Also, without good records of locations, items may be forgotten until they become old and unusable and create a hazard. The compatibility of items in storage areas must also be taken into consideration.

Federal, state, and institutional safety regulations concerning the storing and handling of different types of chemical must be taken into consideration prior to purchasing to prevent violations. The type of violation most often found is flammables in both laboratories and storage areas.

Ability of personnel to handle certain size containers and the economic feasibility of purchasing items in large quantities are two areas so closely related that they are difficult to separate. It is not economical to purchase large containers of items like mineral acids if they cannot be transferred without endangering the health and safety of personnel, nor is it wise to purchase large containers of chemicals that are known to become dangerous as a result of peroxide crystal formation. These create a hazard to existing structures as well as personnel. Anyone who has had to dispose of a drum of ethyl ether is well aware of this fact.

Purchasing in large containers will often dictate the additional purchase of containers for transfer, plus other materials such as pumps, funnels, and related safety items. Work-hour costs and the costs of necessary equipment may eliminate any savings derived from purchasing in large quantities. This method also denies the end user the pertinent information contained on the manufacturer's label, such as handling instructions and first aid information.

The amount of some items being purchased may increase as in the case of a large container being removed from the stock room and located in one laboratory, necessitating additional purchases of the same chemical for other laboratories. A good example of this is bromine, purchased in a 6-lb. bottle and taken to a laboratory that needs only 2 ounces.

There are many different methods of ordering chemicals, and your own procedure will indicate to you which is best for your particular operation.

The method most preferred is a rapid order and delivery system, which often requires the use of open orders or blanket orders to designated suppliers. This allows the purchasing agent to place an order without creating a new purchase order or getting administrative approval for each order, thereby saving time and money.

The advantage of this method is that it allows you to utilize the supplier's storage area until the item is needed at your location. In addition, you receive newer stock on items that tend to deteriorate on the shelf, and your inventory is maintained at a lower level, reducing the amount of investment and the amount of hazardous material on hand in case of an accident. Depending on your location, the materials can be delivered on the same day or within a few days.

The disadvantage is that you will seldom receive the maximum discount that you would receive from buying in case lots or large shipments, and often you must pay freight on small shipments. Also, control of expenditures is transferred from management to purchasing, sometimes creating a situation where unauthorized purchasing can exist.

Ordering to replace items as used, on a daily or weekly basis, requires a system that monitors receiving, issues, on order, and on back order. In most cases for this type of system to function, stock levels must be set and adhered to. The disadvantage of this type of purchasing is that it requires more work-hours than other systems to make it operate properly, and any changes in usage can affect the set stock levels. On the other hand, for large-volume users, this method allows the purchaser to allocate less storage space for items as well as maintain a smaller inventory.

Another method preferred by large users is a contract method. Yearly requirements are contracted for at a certain price, to be shipped at designated times during a given period or when requested. This again allows users to use storage area of the suppliers, but it also commits users to

TABLE A2.1   Purchasing Checklist

Item to be ordered _____

Amount used past _____ months _____

Amount used past _____ days _____

Projected requirements _____
    Based on _____

Proper storage space available   Yes _____   No _____
    Description _____

Equipment on hand to dispense safely   Yes _____   No _____
        (If no, use additional page to list equipment needed and cost)

Qualified personnel to dispense   Yes _____   No _____

Price per _____ in small quantity _____

Price per _____ in bulk quantity _____

Amount of savings _____
    Other advantages, if any _____

Recommended purchasing procedure _____

Recommendations and approval _____
Recommendations and approval _____
Recommendations and approval _____

Action taken _____ Date _____

receive all materials for which they contract, even if these chemicals are no longer needed at that facility.

A third method is an annual or semiannual order. This will normally receive the best discount available but requires the most planning, taking into consideration the following seven items: (a) previous usage; (b) projected usage; (c) available storage space; (d) safety regulations; (e) ability of personnel involved; (f) economics of bulk buying; and (g) alternative purchasing procedures. A combination of any of these methods may work well for some users, depending on location and the volume of chemicals and supplies used (Table A2.1).

Planning for purchasing is essential to the operation of any facility that uses chemicals as a result of the increased cost of chemicals, the necessity of safety in the storage and handling of chemicals, and the ever-increasing regulations and cost of disposal of these items. So, with good data, good storage, trained individuals, and good planning, chemical consuming facilities may be operated wisely and safely.

APPENDIX **3**

A GLOSSARY OF WORD
PROCESSING
AND MICROCOMPUTING TERMS

ALLEN G. MACENSKI

Hughes Aircraft Company
El Segundo, California

ACCESS TIME: The time required to fetch a word from memory.

ACCUMULATOR: An 8-, 16-, or 32-bit register on the CPU chip that serves as workspace for arithmetic, logical, and input/output(I/O) operations. Data fetched from memory can be added, compared, tested, or otherwise operated on and the result held in the accumulator. Programmed transfers of data also pass through it.

ADA: A large-scale computer language commissioned by the Department of Defense, Ada is expected to be the workhorse language of the next two decades. Like COBOL, Ada generally uses full English words; like Pascal, it is designed for modular program construction. Ada is named after Lord Byron's daughter, a pioneer computer programmer.

ADDRESS: A binary number, or symbol for it, that identifies a register, cell of storage, or some other data source or destination. Eight-bit microprocessors usually use 16-bit addresses. Since there are 2^{16} or 65,536 different possible combinations of 16 bits, this allows the direct addressing of 64K (64,000) locations. Sixteen-bit microprocessors use 20-bit (Intel 8086) or 24-bit (Motorola M6800) addresses that can refer to 1-million and 16-million memory locations, respectively.

ALGORITHM: A step-by-step process for the resolution of a problem. Usually developed in outline or as a flowchart before coding, that is, setting it into computer language.

ALPHANUMERIC: The set of all alphabetic and numeric characters, along with related symbols.

ANALOG: Refers to data in the form of continuously changing physical quantities—waves—or devices that operate on it.

APL: A high-level language pioneered by Kenneth Iverson that uses unique keyboard symbols and specially developed mathematical operators.

ARITHMETIC LOGIC UNIT (ALU):   The element in a computer that can perform the basic data manipulations in the central processor.

ARRAY COMPUTER: A computer in which the microprocessors are wired together into an "array" of so many rows and columns. The array computer is extremely fast for two reasons: (a) it can process several commands at the same time; and (b) it can begin processing new information at the same time that it is processing old information.

ARTIFICIAL INTELLIGENCE (AI):   Approximation by a computer and its software of certain functions of human intelligence, learning, adapting, reasoning, self-correcting, and automatic improvement. It can change its own program to adapt to new data it encounters. It is expected to be an important component of the fifth-generation computers of the 1990s.

ASCII: American standard code for information interchange. This is a character code used for representing information.

ASSEMBLER: Program that takes the mnemonic form of the assembly language and converts it into binary object code for execution.

ASSEMBLY LANGUAGE: Developed in the early 1950s so that programmers could use abbreviated word commands (mnemonics) in place of the confusing strings of ones and zeros that are the direct representation of a computer's language.

BANDWIDTH: The maximum number of data units that can be transferred along a channel per second.

BATCH: A processing mode whereby a program is submitted to the computer and the results are delivered back. No interactive communication between program and user is possible.

BAUD: The rate at which data are transmitted over a serial link, such as a telephone line, in bits per second. The format for data transmission is 10 or 11 bits per character, so 300 baud is about 30 characters per second.

BIT: The contraction of "binary digit." A bit always has the value "zero" or "one." Bits are universally used in electronic systems to encode information, orders (instructions), and data Bits are usually grouped in nybbles (4), bytes (8), or larger units.

BOOTSTRAP: A program used for starting the computer, usually by clearing the memory, setting up devices, and loading the operating system from input/output internal or external memory.

BUBBLE MEMORY: A memory device placed on a chip The bubbles are microscopically small, magnetized "domains" that can be moved across a thin magnetic film by a magnetic field. Magnetic-bubble chips that store a million bits of information have been fabricated. Magnetic bubble memory is not as fast as RAM or ROM, but many times faster than mass memory devices such as tapes and disks. Yet like tapes and disks, the information stored inside the magnetized bubbles is retained even after the computer's power is switched off.

BYTE: A group of 8 bits treated as a unit. Under the ASCII format, 7 of the bits provide 128 different codes representing characters, and the eighth is a paritybit for error checking.

CCD: Charge-coupled device. A mass storage device for information on a single chip.

CODING: Putting an algorithm into computer language.

COMPILER: a translation program that converts high-level instructions into a set of binary instructions for execution. Each high-level language requires a compiler or an interpreter. A compiler translates the complete program, which is then executed. Every change in the program requires a complete recompilation.

CPU: Central processing unit. The computer module in charge of fetching, decoding, and executing instructions.

DATA SECURITY: Protection of computerized information by various means, including cryptography, locks, identification cards and badges, restricted access to the computer, passwords, physical and electronic back-up copies of the data, and so on.

DIAGNOSTIC: A program or routine used to diagnose system malfunctions.

DIGITAL: Refers to data in the form of discrete units—"on/off" or "high/low" states—and to devices operating on such data.

DIP: Dual in-line package. A standard integrated circuit (IC) package with two rows of pins at 0.1-inch interval.

DISABLE: To render a control or process temporarily inoperable while another process takes place.

DISK DRIVE: The machinery that contains, rotates, writes onto, and reads from a disk.

DOCUMENTATION: Instructions and other explanatory materials supporting computer hardware and software.

DUMP: To copy the contents of memory or disk to video display, printer, or storage device so that it can be checked out in detail or kept as backup.

EXPANSION INTERFACE: A device to expand the functional capacity of a computer by containing additional memory or controlling more peripherals.

EXTERNAL MEMORY: Used to store programs and information that would otherwise be lost if the computer was turned off. Cassette tapes, disks, bubble memory, and CCDs (charge-coupled devices) are also known as *mass memory* and *removable memory.*

FILE: A logical block of information, designated by name, and considered as a unit by a user. A file may be physically divided into smaller records.

FIRMWARE: A program permanently fixed onto a memory chip (ROM), that is, software on a hardware support.

FLAT PANEL DISPLAYS: Thin, light, with low-power consumption, free from geometric distortion, these displays will begin replacing CRTs in some microcomputers this year. Most familiar is the passive (non-light-emitting) LCD (liquid crystal display). Active (light-emitting) technologies include gas plasma and thin-film electroluminescence (EL). Electroluminescence seems the most likely to develop into a replacement for the CRT in terms of response time, gray scale, and resolution.

FLOW CHART: Symbolic representation of a program sequence. Boxes represent orders of computations. Diamonds represent tests and decisions (branches). A flow chart is the recommended step between algorithm specification and program writing. It greatly facilitates understanding and debugging by breaking down the program into logical, sequential modules.

FORTH: a very efficient, high-level, interpreted language especially suited to microcomputers used in conjunction with sensors and control systems. FORTH's basic commands, very close to assembly language, can be used to define new procedures made by the user. It is called a "threaded language" because these new procedures are created by "threading together" old ones.

GRID: A division of the computer screen into evenly spaced horizontal and vertical lines. Used for locating points on the screen. These points are expressed as row and column coordinates.

HASHING: A system for verifying data input.

HEURISTIC METHODS: Methods that serve to guide or reveal answers that cannot be proved.

HIGH-LEVEL LANGUAGE (HLL):   Problem-oriented languages, much easier to use than machine-oriented languages. They are faster to write than assembly language but produce less efficient object code. COBOL, FORTRAN, PL/1, RPG, and Ada are high-level languages usually compiled. BASIC, Pascal, LISP, and FORTH are usually interpreted. One high-level language statement can produce 10 machine-language statements.

I/O: the communication of information to and from a computer or peripheral device.

INPUT DEVICE: Any machine that allows commands or information to be entered into the computer's main (RAM) memory. An input device could be a typewriter keyboard, an organ keyboard, a tape drive, a disk drive, a microphone, a light pen, a digitizer, or an electronic sensor.

INTEGRATED CIRCUIT (IC):   A complex, microscopic circuit on a chip of silicon.

INTERFACE: The hardware or software required to interconnect a device to a system. One-chip interfaces now exist for most peripherals.

ISAM: Indexed sequential access method. A method for organizing data files for rapid access.

KILOBYTE: "Kilo" means "1000," but "kilobyte" means, precisely, 1024 bytes.

LANGUAGE: In relation to computers, any unified, related set of commands or instructions that the computer can accept. Low-level languages are difficult to use but closely resemble the fundamental operations of the computer. High-level languages resemble English.

LIGHT PEN: An input device for CRT. It records the emission of light at the point of contact with the screen. The timing relationship to the beginning of a scan tells the computer its position on the screen.

LISP: List processing language. A high-level, interpreted language especially effective at recursion and the manipulation of symbolic strings. Good for writing languages, including itself, and for the development of software for artificial intelligence.

LOAD: The action of transferring data in a register or memory location, or a program into a memory area.

LOCATION: The physical place in the computer's memory, with a unique address, where an item of information is stored.

LOGIC: The term used to designate that part of the computer's circuitry that makes logical decisions.

LOGICAL DECISION: The capability of the computer to decide whether one quantity is greater than, equal to, or less than another quantity and then use the outcome of that decision as a cue to proceed in a given way with a program.

LOGICAL OPERATOR: Symbols used in programming to represent the operations of logic, including AND, OR, and NOT. Expressions containing these symbols are often called *Boolean expressions*. (After George Boole, a nineteenth-century British mathematician and logician.)

LOGO: A computer language, developed by scientists at MIT's Artificial Intelligence Laboratory, that has been used experimentally over the past several years in a children's learning laboratory as a tool to help youngsters master concepts in mathematics, art, and science.

LOOP: A group of instructions that may be executed recursively.

LSI: Large-scale integration.

MACHINE LANGUAGE: Set of binary codes, representing the instructions that can be directly executed by a processor.

MAINFRAME: The box that houses the computer's main memory and logic components—its CPU, RAM, ROM, I/O interface circuitry, and so on. The word is also used to distinguish the very large computer from the minicomputer or microcomputer; it usually uses 32-bit term.

MAIN MEMORY: The internal memory of the computer contained in its circuitry, as opposed to peripheral memory (tapes, disks).

MAINTENANCE: The adjustment of an existing program to allow acceptance of new tasks or conditions (e.g., a new category of payroll deduction).

MASK: A pattern, usually "printed" on glass, used to define areas of the chip on the wafer for production purposes.

MASS MEMORY: See external memory.

MENU: A list of programs or applications that are available by making a selection. For example, a small home computer might display the following menu: "Do you want to" (a) "balance checkbook," (b) "see appointments for May," or (c) "see a recipe? Type number desired."

MERGE: A computerized process whereby two or more files are brought together by a common attribute, as zip codes in ascending numerical order.

MICROCOMPUTER: A small but complete computer system, including CPU, memory, I/O interfaces, and power supply. Generally uses 8-bit word.

MICROPROCESSOR: Large-scale integration implementation of a complete processor (ALU + control unit) on a single chip.

MINICOMPUTER: A small computer, intermediate in size between a microcomputer and a large computer. Uses 16-bit word.

MODEL: A computer reproduction (or simulation) of a real or imaginary person, process, place, or thing. Models can be simple or complex; artistic, educational, or entertaining; serious or part of a game.

MODEM: Doculator–demodulator. A device that transforms a computer's electrical pulses into audible tones for transmission over a telephone line to another computer. A modem also receives incoming tones and transforms them into binary impulses that can be processed and stored by the computer.

MULTIPLEXING: Transmitting several signals simultaneously over one data channel in a communications system. Some microprocessors multiplex addresses and data to memory; others use separate address and data channels.

MULTIPROCESSING: A small computer that has more than one CPU and is thus able to process several instructions simultaneously; another form of parallel processing.

NANOSECOND: A billionth of a second. Most computers have a cycle time, or "heartbeat," of hundreds of nanoseconds. Most high-speed computers have a cycle time of around 50 nanoseconds.

NATURAL LANGUAGE: A spoken, human language such as English, Spanish, Arabic, or Chinese. In the future, small computers will probably be fast enough and have large enough vocabularies to enable one to talk to them, using one's native language. Yet there will still be a need for special computer languages that are more efficient than native languages for handling certain kinds of tasks.

NETWORK: Several intelligent devices such as microcomputers interconnected so that they can send instructions and data back and forth between themselves, thus forming a larger computational system.

NYBBLE: Usually 4 bits, or half of a byte.

OBJECT PROGRAM: A program in machine-readable form. A compiler translates a source program into an object program.

OCR: Optical character recognition. The data input system that allows the computer to "read" printed information.

ON LINE: Directly connected to the computer system and in performance-ready condition.

OPERATING SYSTEM (OS):   Software required to manage the hardware and logical resources of a system, including scheduling and file management.

OPTICAL WAND: An input device with a photoelectric "camera" that senses black and white light patterns.

OUTPUT DEVICE: A machine that transfers programs or information from the computer to some other medium. Examples of output devices include tape disk and bubble memory-drives; computer printers, typewriters, and plotters; the computer picture screen (video monitor); robots; and sound synthesis devices that enable the computer to talk and play music.

PASCAL: A high-level language developed to teach structured programming. Pascal has become popular for general use.

PCB: Printed circuit board.

PERIPHERAL: Any human interface device connected to a computer.

PICOSECOND: A trillionth of a second. Even light, the fastest substance in the universe, can travel only one one-hundredth of an inch in a trillionth of a second.

PILOT: An interpreter language, like BASIC and FORTH. PILOT is a simple programming language that is ideally suited for use in computer-aided instruction applications.

PIPELINE COMPUTER: A computer with a string of "processor" CPUs all capable of executing an operation simultaneously.

PIXEL: The computer picture screen is divided into rows and columns of tiny dots, squares, or cells. Each of these is a pixel (from picture element).

PL/1: A high-level, compiled language developed by IBM during the 1960s that combines the best features of COBOL and FORTRAN and generates efficient object code. Its large vocabulary and large compiler were only recently adapted for 8-bit microcomputers.

PLASMA-RAY DEVICE: A flat computer picture screen based on a gird of metallic conductors separated by a thin layer of gas. When a signal is generated at any intersection along the grid, the gas discharges and causes the transparent screen to glow at this point. In the future, the thinner, more reliable, plasma ray and solidstate devices will replace CRTs as computer video display terminals.

PLOTTER: A mechanical device for drawing lines under computer control.

PORTABILITY: The property of software that permits its use in a variety of computer environments.

PROGRAM: A sequence of instructions that results in the execution of an algorithm. Programs are essentially written at three levels: (a) binary (can be directly executed by the MPU); (b) assembly language (symbolic representation of the binary); and (c) high-level language (e.g., BASIC), requiring a compiler or interpreter.

RAM: Random access memory. Denotes in fact addressable Read/Write LSI memory.

RANDOM ACCESS: An access method whereby each word can be retrieved directly by its address.

READ: To accept data from a disk, card, and so on, for storage and/or processing.

REAL TIME: Immediate and concurrent processing.

RECORD: The process of writing information, or the block of information itself.

REGISTER: One-word memory, usually implemented in fast flip-flops, directly accessible to a procesor. The CPU includes a set of integral registers that can be accessed much faster than the main memory.

RESOLUTION: The quality of the image on the CRT, as influenced by the number of pixels on the screen. The greater the number of pixels, the higher the resolution.

ROBOT: Any stored-program device capable of altering its external environment.

ROM: Read-only memory.

SENSOR: Any device that acts as "eyes" or "ears" for a small computer. Types of sensor include photoelectric sensors that are sensitive to light; image sensor cameras that record visual images and transform the images into digital signals; pressure sensors that are sensitive to any kind of pressure; contact sensors that record IR information; and ultrasonic transducers that produce a high-frequency sound wave that bounces off objects and lets the computer calculate the distance between itself and those objects.

SEQUENTIAL ACCESS: The method in which data are accessed by scanning blocks or records sequentially.

SERVICE BUREAU: A company that offers time-sharing, programming, and other computer services to businesses.

SIMULATION: A computerized reproduction, image, or replica of a situation or set of conditions.

SMALLTALK: A computer language developed by the Xerox Palo Alto Research Center. When it finally becomes commercially available, Smalltalk may prove to be the most powerful language for small computers of the future.

SOFT COPY: Information contained magnetically in storage.

SORT: A full or part program to reorder data sequentially, usually in alphabetic or numeric order.

SOURCE PROGRAM: A program written in a language that is not directly readable by a computer; it must be converted to an object program for use by a computer.

STATIC RAM: Unlike ordinary, volatile memory, static memory retains its contents even when the main current is turned off. The trickle of electricity from a battery is sufficient to refresh it.

STRING: A linear series of symbols treated as a unit, such as this sentence.

STRING FLOPPY: An endless loop of recording tape in a cartridge used as external memory.

STRUCTURED PROGRAMMING: In a program, proceeding in a systematic way from section to section rather than branching widely on GOTO in instructions.

SUBROUTINE: A programmed module for a special task (e.g., computing a square root) that can be called in at any point of the main program.

TAPE: Inexpensive mass storage medium. Must be accessed sequentially. Convenient for large files.

TELEPRESENCE: By using all of a robot's sensors, a person can be electronically aware of the robot's immediate environment and control the robot's actions just as if the person were actually in the location of the robot.

TELESCREEN: A two-way, audiovisual "television" used to monitor and control remote activities.

TELETEXT: Textual information transmitted to people's homes through television. Information is usually maintained and updated on a computer. A teletext system often allows two-way interaction by creating a viewer–computer link over a telephone line.

THROUGHPUT: The number of instructions executed or the amount of data transmitted per second. A measure of a computer's power and efficiency; 10 MIPS means 10 million instructions per second.

TIME SHARING: Occurs when a single computer has multiple users who are each getting a "slice" of each second of the computer's processing time.

TRANSPORTABLE: Of software: usable on many different computers.

UTILITY: Software for routine tasks or for assisting programmers.

VIDEODISK: A recordlike device storing a large amount of audio and visual information that can be linked to a computer A single side of one videodisk can store the pictures and sound for 54,000 separate television screens. Yet any one of these screens images can be accessed and displayed.

VIRTUAL MEMORY: A major extension of main memory address space into a secondary memory such as hard disk. A program in virtual memory is divided into segments called *pages*, and these pages are read into main memory as needed.

VLSI: Very large scale integration. In practice, the compression of more than 10,000 transistors on a single chip.

WAFER: Three- or four-inch slice of silicon that is overlaid with microscopic circuitry and broken up into many individual computer chips.

WORD: The unit of information stored in a computer's memory, moved in parallel along its data paths, and worked on in its registers. Word size is an important distinction between classes of computer. Small microprocessors have a 4-bit (one nybble, or half-byte) word. The classic microcomputer has an i-bit (one byte) word, although micros with 16-bit words are not appearing. Minicomputers usually use 16-bit words. Mainframes use words of 32 bits or more in some cases.

WRITE: To transfer data from internal to external memory.

APPENDIX 4

SAFETY EQUIPMENT FOR STORAGE OF FLAMMABLE LIQUIDS

Fire-protected facilities are essential for the safe storage of flammable liquids. Chapter 1 gives recommendations for storeroom construction and capacity for storing large amounts of flammable liquids; however, there will still be other areas where limited quantities are kept. This appendix briefly identifies and discusses *safety cans, flammable liquid storage cabinets,* and *especially designed refrigerators* for the purposes of limited-quantity storage.

Safety cans are containers that have built-in safety features for protecting flammable liquids from exposure to a fire situation. In a fire situation, a safety can is exposed to extremely high tempertures. This heat is trans-

FIGURE A4.1   Cutaway view of safety can reveals position of cylindrical flame arrestor. (Photograph courtesy of The Justrite Manufacturing Company, Des Plaines, Illinois.)

mitted to the contents, which, in turn, boil and produce a large vapor pressure. Every safety can is fitted with a spring-loaded cap that vents these vapors safely without bursting the safety can. The other safety feature of a safety can is the flame arrestor, which is a cylindrical wire screen. Vapors emitted from a safety can will ignite when exposed to the flames of a fire. Since flames usually flash back to the source of liquid, the flame arrestor serves as a heat dissipator. The temperature in the space above the liquids in a safety can is lowered below the ignition temperature, and ignition of the contents is eliminated. Figure A4.1 shows a cutaway view of a safety can that reveals the positioning of a flame arrestor.

Safety cans are available in several sizes, the largest of which has a 5-gallon capacity. Table H-12 in OSHA regulation No. 29 CFR 1910:106 shows the maximum allowable size of safety cans for the various classes of flammable and combustible liquid. Occupational Safety and Health Administration Code 1910.106 (d)(2) requires that only *approved* containers (except those in DOT-Spec metal drums) to be used to store

FIGURE A4.2 A metal storage cabinet approved for storing flammable liquids. Such cabinets are constructed with double metal walls separated by a 1½-inch air space (Photograph courtesy of Lab Safety Supply Company, Janesville, Wisconsin.)

flammable liquids and combustibles. Many safety cans commercially available have been tested and approved by Underwriters Laboratories (UL) and/or Factory Mutual System (FM). In addition, OSHA regulations [29 CFR 1910.144 (a)(ii)] require safety cans to be painted red and carry either a yellow band or have the name of the contents painted or stenciled in yellow on the can for flammable liquids with a flash point at or below 80°F.

Flammable liquid storage cabinets are designed to keep the temperature at the top center of the cabinet interior below 325°F when subjected to a

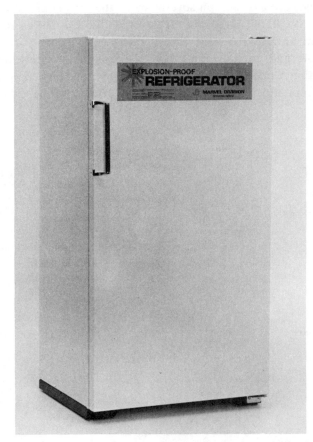

FIGURE A4.3   An explosionproof refrigerator. To ensure maximum shielding of an "explosion-prone" area from ignition sources, this refrigerator must be wired directly into conduit. (Photograph courtesy of Marvel Division, Dayton-Walther Corporation.)

10-minute fire test. (*Note:*A storage cabinet for flammable liquids is *not* fireproof, but only protects the contents from extreme temperatures for a limited time to allow evacuation of personnel and prompt entry of fire fighters.)

Cabinets built to withstand the temperature rating during the 10-minute fire test (prescribed in NFPA 251-1969) are acceptable by OSHA standards if (a) the maximum capacity of Class I and II liquids is not more than 60 gallons (or more than 120 gallons for Class II liquids) and (b) the cabinet is labeled with conspicuous lettering "Flammable—Keep Fire Away."

Construction requirements for wooden and metal cabinets acceptable for the storage of flammable and combustible liquids are given in 29 CFR 1910.106 (d)(3). Figure A4.2 shows a picture of a double-wall metal storage cabinet for flammable liquids.

For small quantities of Class IA liquids (which have low boiling points), refrigerated storage may be necessary to prevent volatilization. Special refrigerators that may safely store flammable liquids have a spark-free interior in that all wiring and thermostat controls have been removed from the interior. Two types of these refrigerators are commercially available— a "flammable liquid storage" model and an "explosionproof" model. A "flammable liquid storage" model is normally used in a nonexplosive area where no flammable vapors are present. Such a refrigerator is normally powered through a standard three wire cord plugged into an electrical outlet. An "explosionproof" refrigerator (Figure A4.3) is required when the area in which the refrigerator will be located has the potential for ignition of flammable vapors. An explosionproof refrigerator is supplied with a "pigtail" cord that must be wired directly to a power source using metal conduit as specified by local electrical codes. Choosing the appropriate refrigerator will depend on the area in which it will be located.

APPENDIX **5**

FLASH POINTS OF COMMON FLAMMABLE LIQUIDS

The following are Class IA flammable liquids (flash point <73°F; boiling point <100°F):

Flammable Liquid	Flash Point (°F)
Ethyl chloride	−58
Pentane	−57
Ethyl ether	−49
Acetaldehyde	−36
Isopropylamine	−35
Ethyl formate	−2
Ethylamine	0

The following are Class IB flammable liquids (flash point <73°F, boiling point ≥100°F):

Flammable Liquid	Flash Point (°F)[a]
Naphtha[b]	−40 to 68
Allyl chloride	−25
Carbon disulfide	−22
Isopropyl ether	−18
Acrolein	−15
Hexane	−7
Cyclohexane	−4
Ethyl bromide	<−4
Nickel carbonyl	−4
Acetone	1.4
1,1-Dimethylhydrazine	5
Tetrahydrofuran	6
Butyl amine	10
Benzene	12
Methyl acetate	14
Methyl ethyl ketone	21
Ethyl acetate	24
Heptane	25
Acrylonitrile	30
Butyl mercaptan	35
Toluene	40

Flammable Liquid	Flash Point (°F)[a]
2-Pentanone	45
Methyl methyacrylate	50 (oc)
Methanol	52
Isopropanol	53
Dioxane	54
Ethylene dichloride	55
Octane	56
Propanol	59
Sec butyl acetate	62
Pyridine	68
Allyl alcohol	70
Butyl acetate	72

[a]Closed cup values are given unless where denoted by "oc" (open cup).
[b]Borderline Class IA.

The following are Class IC flammable liquids (Flash point ≥73°F, but less than 100°F):

Flammable Liquid	Flash Point (°F)[a]
Methyl isobutyl ketone	73
2-Butanol	75
n-Amyl acetate	77
2-Hexanone	77
Isoamyl acetate	77
Xylene	81
Butyl alcohol	84
Chlorobenzene	84
p-Ansidine	86
sec-Amyl Acetate	89
Styrene	90
Ethylene diamine	93
Morpholine	95
Turpentine	95

[a]Flash-point values were taken from *NIOSH/OSHA Pocket Guide to Chemical Hazards*, DHEW (NIOSH) Publication No. 78-210, Fourth printing, August 1981.

APPENDIX 6

CHEMICAL STORAGE CHECKLIST

Taking stock of current storage conditions and procedures is the first step in managing a safe chemical storeroom. The following checklist* has been developed to help assess safety in the storeroom. The checklist format not only facilitates a systematic assessment of storage and housekeeping conditions, but also identifies general and specific areas of concern. The completed checklist serves as a record of needed improvements. An affirmative answer to each item indicates a satisfactory storage condition.

Yes   No   NA†

STORAGE AREAS

Storage rooms are properly marked or identified.

Storage areas are secured whenever not in use and are available only to authorized personnel.

Storage areas are free of blind alleys.

Storage areas have two or more clearly marked exits.

Storage areas are well illuminated.

Storage areas are well ventilated, with exhaust air leaving the building. (Beware of recirculating systems.)

Storage areas have adequate air-conditioning and/or dehumidifier systems to provide a cool, dry atmosphere.

Open flames, smoking, and localized heating units are not permitted in chemical storage areas.

Mixing or transfer of chemicals is not allowed in storage areas.

Aisles in the storage area are free from obstruction.

*The Chemical Storage Checklist originally appeared as part of an article: "Safe Storage of Chemicals: A Checklist for Teachers." *The Science Teacher,* **48**, (2) (February 1981). Reprinted with permission from *The Science Teacher* (February 1981), published by the National Science Teachers Association, Washington, DC and Lab Safety Supply Company, Janesville, Wisconsin.
†Not applicable.

Ladders with handrails are available where needed.

Sources of sparks are completely eliminated from the storage area.

SHELF STORAGE

Large bottles and containers are stored on shelves no higher than 2 feet from the floor.

Containers of chemicals are stored below eye level.

Shelves have raised edges or rim guards to prevent the accidental dislodging of containers.

Reagent bottles or containers do not protrude over the shelf edges.

Enough space is available so that chemicals are not overcrowded.

Empty bottles are removed from stockroom shelves.

Shelves are level and stable. Shelving units are securely fastened to wall or floor.

Weight limit of shelves is posted and not exceeded.

Shelves are clean—free of dust and chemical contamination.

STORAGE CONTAINERS

Storage containers are inspected periodically for rust, corrosion, or leakage.

Damaged containers are removed or repaired immediately.

Chemicals are kept in airtight bottles, not in beakers or open vessels.

Stoppers from an airtight seal with containers.

Stoppers are easily removed from bottles or containers.

Containers of mercury are well stoppered.

Carboys are used for storing chemical solutions.

Eye-dropper bottles are not used for storing corrosive or water-reactive chemicals.

All carboy spigots are leaktight and drip free.

Dispensing tubes on carboys are free of corrosion or aging.

LABELING OF CHEMICAL CONTAINERS

National Fire Protection Association (NFPA) hazard labels and identification system are used for labeling all dangerous chemicals.

All containers are clearly labeled as to contents.

Labels are readable and free of encrustation or contamination.

Labels are firmly attached to containers.

Chemical containers are labeled with the appropriate warning (poison, corrosive, etc.).

All container labels include both date of receipt and anticipated disposal.

Labels include precautionary measures for the specific chemical.

HOUSEKEEPING

Cleanliness and order are maintained in the storage areas at all times.

Unlabeled, contaminated, or undesirable chemicals are discarded properly.

Chemicals in storage cabinets and on shelves are inspected for decomposition on a regular basis. An inspection log is kept.

Unused chemicals are never returned to stock bottles.

Packing materials and empty cartons are removed at once from the stockroom.

Waste receptacles are properly marked and easily located.

Separate disposal containers are available for broken glass.

Environmentally safe disposal methods have been arranged for dangerous waste chemicals.

GAS CYLINDERS

All gas cylinders are secured to prevent falling over.

Gas cylinders are stored away from direct or localized heat, open flames, or sparks.

Gas cylinders are stored in a cool, dry place away from corrosive fumes or chemicals.

Gas cylinders are stored away from highly flammable substances.

Empty gas cylinders are labeled "EMPTY" or "MT."

Empty gas cylinders are stored separate from full gas cylinders.

Flammable or toxic gases are stored at or above ground level, never in basements.

Cylinders of incompatible gases are segregated by distance.

When gas cylinders are not in use, the valve cap is securely in place to protect the valve stem and valve.

A hand truck is available for transporting gas cylinders to and from the storage area.

FIRST AID

First aid supplies are readily available and have been approved by a consulting physician.

First aid cabinets are clearly labeled.

Emergency room staff, with medical personnel specifically trained in response to chemical exposure, are readily available.

Blankets are available for shock cases and for protection of the injured.

Supervisors are trained in resuscitation.

Emergency telephone numbers are posted on or near the telephone.

Eyewash and shower facilities are within 10 seconds or 100 feet of the site of hazardous materials.

Eyewash and shower facilities are periodically inspected and maintained.

Hand-washing facilities are readily available for stockroom personnel.

EMERGENCY PREPAREDNESS

An emergency warning system is available in the event of an accident.

Emergency and evacuation procedures are known by stockroom personnel.

Emergency and personnel protective equipment is located *outside* areas where accidents may occur.

At least two self-contained breathing apparatus are available for emergencies. Personnel are trained in usage.

Equipment and supplies for cleaning up spills are readily available.

Fire extinguishers are immediately accessible.

Fire extinguishers are periodically inspected and maintained.

Fire and smoke alarms are located in fire-prone areas, with periodic maintenance and inspection.

CHEMICAL STORAGE

Chemicals are not exposed to direct sunlight or localized heat.

Containers of corrosive chemicals are stored in trays large enough to contain spillage or leakage.

Chemicals are stored by reactive class (i.e., flammables with flammables, oxidizers with oxidizers).

An incompatibility–compatibility guide is available to indicate arrangement of chemicals.

Incompatible chemicals are physically segregated from each other during storage.

Acids

Large bottles of acids are stored on a low shelf or in acid cabinets.

Oxidizing acids are segregated from organic acids, flammable, and combustible materials.

Acids are separated from cautics and from active metals such as sodium, magnesium, and potassium.

Acids are segregated from chemicals which can generate toxic gases on contact, such as sodium cyanide and iron sulfide.

Bottle carriers are used for transporting acid bottles.

Spill control pillows or acid neutralizers are available for acid spills.

Caustics

Caustics are stored away from acids.

Solutions of inorganic hydroxides are stored in polyethylene containers.

Spill control pillows or caustic neutralizers are available for spills.

Flammables

Stockroom personnel are aware of the hazards associated with flammable materials.

All flammable liquids containers are in compliance with the maximum container sizes found in Table H-12, 29 CFR 1910.106.

OSHA/NFPA-specified safety cabinets are used for the storage of flammable liquids.

Flammable liquids are stored in accordance with NFPA Standard No. 30, Flammable and Combustible Liquids Code.

Flammables are kept away from any source of ignition: flames, heat, or sparks.

Approved refrigerators are used for storing highly volatile flammable liquids.

All electrical service equipment is explosionproof for the appropriate class and group of flammable liquids.

Bonding and grounding wires are used where flammables are stored and dispensed.

Peroxide Forming Chemicals

Peroxide-forming chemicals are stored in airtight containers in a dark, cool, and dry place.

Peroxide-forming chemicals are properly disposed of before the date of expected peroxide formation.

Suspicion of peroxide contamination is immediately evaluated by use of safe procedures.

Chemicals are labeled with date received, date opened, and disposal date.

Water-Reactive Chemicals

Chemicals are kept in a cool and dry place.

In case of fire, a Class D fire extinguisher is used.

Oxidizers

Oxidizers are stored away from flammable, combustible, and reducing agents (e g., zinc, alkaline metals).

Toxic Compounds

Toxic compounds are stored according to the nature of the chemical, with appropriate security employed where necessary.

A "Poison Control Network" telephone number is posted.

INDEX

DATE DUE

MR 26 02			